Benjamin Ward Richardson

The Commonhealth

A Series of Essays on Health and Felicity for Every-Day Readers

Benjamin Ward Richardson

The Commonhealth
A Series of Essays on Health and Felicity for Every-Day Readers

ISBN/EAN: 9783337418304

Printed in Europe, USA, Canada, Australia, Japan

Cover: Foto ©berggeist007 / pixelio.de

More available books at **www.hansebooks.com**

THE COMMONHEALTH

A SERIES OF ESSAYS ON HEALTH AND FELICITY

FOR EVERY-DAY READERS

BY

BENJAMIN WARD RICHARDSON, M.D., F.R.S.

FELLOW OF THE ROYAL COLLEGE OF PHYSICIANS
AND HONORARY PHYSICIAN TO THE ROYAL LITERARY FUND

'Homines ad deos nullâ re propius accedunt quam salutem hominibus dando'
CICERO

Men never nearer to the gods attain
Than in the art of giving health to men

LONDON
LONGMANS, GREEN, AND CO.
1887

All rights reserved

TO

GEORGE GODWIN, F.R.S., F.S.A.

My dear Godwin,

In the band of useful and distinguished men who have striven to make this century assume the vesture of health and the joy of life, you, in your quiet and persistent way, have been one of the foremost, your mind ever set on the homes of the people as the centres from which alone the blessings of health can spring. To you, therefore, a pioneer of pioneers, I dedicate this volume, as a slight token of the true respect and affection which during the many long years of our friendship I have felt towards you, and in which I pray to remain

Yours always truly,

BENJAMIN WARD RICHARDSON.

PREFACE.

The twelve essays published in this volume consist chiefly of addresses which on various occasions have been delivered before the sanitary and other societies of this kingdom, and which, having been favourably noticed by the press as current topics, have since been asked for in more permanent form.

25 Manchester Square, London, W.
May 1, 1887.

CONTENTS.

	PAGE
THE SEED-TIME OF HEALTH	1
HEALTH AND RECREATION	34
HEALTH AND RECREATION FOR THE YOUNG	61
HEALTH THROUGH EDUCATION	90
NATIONAL NECESSITIES AS THE BASES OF NATIONAL EDUCATION	115
DISEASES INCIDENT TO PUBLIC LIFE	157
WOMAN AS A SANITARY REFORMER	167
DRESS IN RELATION TO HEALTH	201
THE POVERTY OF WEALTH	233
UPPER AND LOWER LONDON	254
FELICITY AS A SANITARY RESEARCH	272
CYCLING AS A HEALTH PURSUIT—PHYSICAL AND MENTAL	303

THE SEED-TIME OF HEALTH.[1]

In the depths of the night, in a climate where night is short, in the midst of that short interval when even the gods are supposed to rest, when the sun-god himself has withdrawn from the earth, and the sun sees not the deeds of men, some men and women of the earth, in solemn silence, bring something forth from home.

If they should speak there would flow from the lips of those people a language so beautiful, so perfect, so expressive, that though the listening ear were foreign to it and understood it not it would be held listening. But there is not a sound.

If these people could be seen in their fair stature and build of body, draped in their loose garments, the eye, like the ear, would be vanquished. Such incomparable beauty! Should a sculptor want a model for a work he would leave for all time, he would find it in them; should a painter want a face for his perfected art, he would find it in them; should a poet want a theme for a song on living beauty, he would find his inspiration in them; should a physician want a text for a discourse

---

[1] Presidential Address to the Brighton Health Congress. Delivered in the Dome, Brighton, on Tuesday evening, December 13, 1881.

on the types of health and sanity, he would find it in those types of beauty.

In those faces, which actually live to this hour in marble more precious than gold, there would be seen, if they were unveiled from this awful stillness and darkness of the night, two living passions, engraved in life through expression of the soul, resigned grief and sublime fear. What has happened can never be recalled, and grief, therefore, is chastened by reason; but what has happened is so unnatural, so wrong, that reason, in its turn, is sublimed to fear. It is so terrible, none must look on it: if the sun-god, source of light and life, should see it, he might hide his face and punish all the races of mankind.

Well may there sit on every face the chaste beauty of resignation and sublimity of fear!

What can have happened?

There is something that is being carried tenderly, awfully! It is a casket, small and light. It might be a cradle or a cot supporting some object of tender solicitude. A child! Yes, a child, in all its childish wealth, its golden tresses on its pillow, its features divinely fair and spiritual, its limbs the ideal of grace. Surely in the dead of this night it sleeps, and they are taking it to some golden coast, where in the morning it will greet the sun, lave in the azure sea, listen to the shell picked up by the shore for the mysterious music, and bask in pleasure.

Alas! no. As the earth is now dead and silent, its soul of sun withdrawn, so is the soul of that human lovely form; and as the earth is proceeding to enter once more the eternal fire that at once animates it and destroys it, so this child of earth is being carried to the pyre.

Beyond expression terrible this event, that they, the bearers and followers, should be so ignorant as to let such beaming beauty die. Had it lived its course, played its mortal part, and like the ripe grain fallen fairly under the sickle of the immortal reaper, then, though a thousand suns had shone, the event had been natural, honourable. Then this ceremonial had been public as the day. Tears might have moistened the eyes of the lookers on, but there would be no shame; the deeds of the dead might be themes of honour, or fame, or joy; but shame, no trace of it. The shame is now; the shame that must be hidden in darkness of darkness, as a crime against knowledge, and love, and family, and country, and time! the shame that life in its earliest dawn should be let go, and run no Olympian game, and sing no song, and tell no history, and plant no work of art, and hold no standard, and fulfil no task of duty. They veil themselves from the truth that they may awake us from a deathly-dream. Let them pass from us also as a dream. Yet the dream is true, for I have embodied in these sentences an idea of mankind in that period of human history when, as by a miracle, the human soul burst into the flame which to this day is our great source of intellectual light; the flame that in its own home went out, but from which, while it burned, all the world lighted a torch and carried it away.

While the sculptor of to-day still strikes a light from the dead of that period of intellectual glory, from the very marble into which its fervid life was infused for ever, let us who deal with actual life strike a light from the sentiment regarding the young who fell as they were rising from the drowsy torpor of infancy into the waking dreams of adolescence, instead of passing, in natural course, through manhood or womanhood, towards maturity, towards drowsy decline.

These wise people knew that life ought to be a perpetual feast. They not only knew the fact, they acted up to it. They were equally well aware that a long and perfect life could alone be attained by perfection of life at its opening, in the seed-time of health. To die at that time was, therefore, an offence against natural rule, against reason, against sentiment. The knowledge of such an event was death to the brain, death to the heart. In this seed-time of health the life was to be made, the life that was to be in truth a life worth living. Animals beneath men, that are worthy of going through their appointed time, and of being made both useful and beautiful, must have their seed-time of health. Shall their human masters be less cared for? If the masters are to be mere slaves, yes; and then it were a pity and a danger; for they who have no respect for life and beauty, who drag through existence and grow weary of it, are to be trusted neither with life, beauty, nor fame.

In the history of great truths derived from the Hellenic wise times, there is not one truth so great as this, and not one so completely missed. It is the secret that was lost. In our day we have lost it so severely that it might never have been in existence for aught we seem to care. The key to all we would have, the key to the gates of health and happiness, has been lost as if it had never been found.

In point of health our children in these times, proud

as we are of these times, are our reproach. Where is there a healthy child ? I have never seen one. I might search through the length and breadth of the island, I could not find one. You may put before me a child in all its innocence. It has done no wrong that it should suffer; it may show to the unskilled mind no trace of disease; and yet I know that if I or any skilled observer will look into the history of the life in question, it cannot be found intrinsically sound. It will have to battle with future dangers sufficient for the soundest to meet; but it is not itself free from dangers other than those that are prospective and avoidable. It is sure to have some inherited failures, and too likely some that will help to increase the independent risks that lie before it.

So our children under five years are expected to die in what may almost be called a definite proportion. He is a fortunate man who, having four children born to him, retains three alive. Later on, for a short time, the danger is reduced; with adolescence it recurs. Again it retreats, but with such failure all along the line, that one-third of the allotted life, the life that would be were it planted in sound health, is only attained.

And for this we have no shame. The sun, the moon, and all the stars may witness our miseries, and we may grieve, but we have no shame. There is an assembly of learned men which I sometimes visit, an assembly of earnest men who are bent on understanding to the full these human failures from health. These men spare no pains, and to gain a spark of light will labour like miners in a mine. When last I visited them a puny feeble spark of life was in their presence undergoing their searching yet kindly scrutiny. Except that it cried a little and laughed a little in changing mood, this spark of life might have been considered a pathological specimen, and in truth it was discussed as such. No one there had a thought of that small life developing into wholesome life and passing through its natural term; not one was there who did not know that the continuance of life was impossible, and that nothing could be done to save the life. The intent was to study the pathology, and fix that by name. They said, when their technical language was translated, this child is suffering from the error, some would say the sin, of its parents. How deep did this error go? In what strange forms did it appear? How singular that the nervous system, once impressed with the poison of that error, should impress another nervous system, and so modify the nutrition of the organism to which it belonged as to cause false nutrition of internal organs and of the very bones themselves! In a whisper one of the learned expressed to another one the pity 'that such a specimen of humanity should ever have been born, to breathe and take notice, and smile, and cry, and love, and suffer, and die, and we able to do nothing for it except hope for the relief that should end in the earliest death.'

I belong to a committee which takes under its care another class of sad childhood. The members of this community pass before us deaf and mute. We try to give them the powers of intelligent converse by laborious and artificial means, and we do some good; but the train of sufferers passes by, and we know that full half are mute from the undeveloped brain; that they are practically lost to life. It is not that the one sense is lost, and thereby the means of expression by intelligible language; it is not even that the nervous organisation which ministers to intelligence is low; it is that these deficiencies are some of the outward signs of a general deterioration of body, and that there is scarcely a structure which the eye of science would recognise as moulded in health.

Passing from the sphere of general observation, from modified to destroyed vitality, I find more startling facts at hand. A short unpretending essay reached me not long ago in which the writer, who in his too great modesty conceals his name, epitomises the facts he has collected respecting the attainment of maturity in peoples of different nations. He tells us that of ten children born in Norway a little over seven reach their twentieth year; that in England and in the United States of America somewhat less than seven reach that stage; that in France only five reach it; and in Ireland less than five. He tells us that in Norway out of ten thousand born rather more than one out of three reaches the age of seventy; in England one out of four; in the United States, if both sexes be computed, less than one out of four; in France less than one out of eight; and in Ireland less than one out of eleven. And he adds this significant computation, based on what may be called the commercial view of the vital question. 'In producing dead machinery the cost of all that is broken in the making is charged to the cost of that which is completed. If we estimate by this same rule the cost of rearing children to manhood, if we calculate up the number of years lived by those who have fallen, with the years of those who have passed successfully to manhood, there would be found between the two extremes presented in Norway and Ireland—both, be it observed, unnatural—a loss of one hundred and twenty per cent. greater in the first year of life, seventy-five per cent. greater in the first four years of life, and one hundred and twenty per cent. greater in the years between the fifth and the twentieth, in Ireland than in Norway. In Norway the average length of life of the effective population is thirty-nine and rather more than a half years; in England, thirty-five and a half years; in France, not quite thirty-three years; and in Ireland, not quite twenty-nine years. Thus, again comparing the best with the worst of a scale of vitality in which both are bad, in Norway the proportion of the population that reaches twenty survives nearly forty years, or four-fifths of the effective period, to contribute to the wealth of the community; while in Ireland the same proportion survives less than twenty-nine years, or considerably under three-fifths of the effective period.

When we are sitting in the family circle and are speaking of families that lie within our cognisance, we estimate in the most natural way the happiness of the families by the health they represent. We may thoughtlessly speak of other standards of measurement. We may for a moment dwell on the riches of the house; on the luxuries that are to be seen in it; on the influence which the owners of it might or do exercise in the social sphere, and such like sentiments. But, after all, these rest on health as the basis of the happiness. If one out of every two of the offspring of the favoured house have died, if some who have not died are mute to the world or otherwise stricken, we soon fall into more thoughtful mood, and say that even this rich home is not a possible home for happy life. Pleasures there may be, happiness there cannot be.

How much worse the estimate of a family in which, together with the vital failures, there is the lack of all that is necessary to make the burden of life endurable. The favoured in health and means wonder, when they think of it, how such unfavoured endure the life they live. In that sentiment no maudlin canker lies; it is as hard and as free from poetry as a mathematical problem; and is, for that reason, a sentiment which, above every other, is persistently preserved.

What is true of family circles is equally true of nations. Rest, quiet of nations, repose for cultivation of refined arts and sciences, happiness derived from healthy and vigorous minds and intended for healthy, vigorous, and wholesome purposes, there cannot be, when one in two of life can only reach maturity with a survival of three-fifths of effective population. Such a national family presents persistent mourning. It sits for ever in gloom; the blinds of its home are always drawn. It broods; it attributes, as all heart-stricken mourners do, the loss it has sustained to every imaginable and unimaginable cause. It thinks with incoherency; speaks now with hysteric grief, then with hysteric rage, and acts the same. In a word, it follows natural law. State physicians tender their remedies for such families of nations and call themselves curers, as if that could be cured which is Nature pursuing her merciless course towards her merciful dispensations, in correction of those who have outraged her.

I have named this discourse 'The Seed-time of Health,' and in the sentences foregone I have tried to strike a contrast, and thereby to give to sanitation a broader meaning as a practical science than is commonly connected with it as a system of details respecting ventilations, sewer traps, and the like.

I want to point to health as the all-in-all to man; the gate of life, leading to the truly good in politics, art, science, letters—ay, and religion, not less than the least of everything. The strain of my argument is, that, unless we make the early life of our children a seed-time of health; unless we, from the root of life, so change the conditions which now exist, all our other measures are practically valueless.

At this moment we have not, as a nation, got this notion set in our minds in such degree as even to accept it, basic as it is, as worthy of serious thought. We have no shame when our young fail and die. Grief we have, fond memories we have; but shame, none. We bury our young as if the act were natural, and erect memorials of it. We print obituaries of the young dead; we read the terrible obituaries of the Registrar-General; we discuss in congresses like these the cost of young life; but the shame of the Greek touches us not. The knowledge of the troubles which flow from the lack of the shame reaches us not.

One bright Sunday morning I was in Dublin, in the Phœnix Park. A great crowd formed a vast ring, to the borders of which I made my way. A wrestling match! Men of different counties wrestling in deathly earnest, the lookers-on intent to terror. On not a face in that multitude, barring the faces of some four or five cockneys, who had a car all to themselves, and grinned as foolishly as they chattered and chaffed, was there so much as a smile; the victors were approved, but not cheered. If this be sport, I felt, it is the strangest I ever knew since I read of Christian trying to be merry in the castle of Giant Despair. In that same day I traversed the city to see authority armed to the teeth in utterly joyless open places. I visited an exhibition of pictures to experience the same sense of all-pervading oppression. I followed a crowd, and found myself one of another multitude going out of the city until we reached a place where the members of that multitude were burying their dead; and, as they swept by, the train of young dead that was carried in the sight of the sun to sleep in that resting-place was to me as appalling as it was revealing. It was like lightning in persistent discharge. Peace, progress, content, happiness, with this discharge of fearful facts in view! A fable! 'As is the earthy,' says the priest, 'such are they also that are earthy;' and I knew that I had never understood the saying before.

It struck me for the first time, as I witnessed this painful phenomenon, that, with so much young death, there could no more be health in the body politic than in the body corporeal. We sanitarians are, however, only bound to treat of that which belongs to our own labours; and, acknowledging the perils incident to early life, and it may be even recognising the shame of them, have before us the question of their prevention from its health side alone.

That we may approach this task with intelligence, let us for a short time glance at the nature of the perils which beset the springtide of human life, and the period bounded by maturity.

The perils are of four kinds:—

1. Those that are inherited.
2. Those that are accidental.
3. Those that are inflicted.
4. Those that are acquired.

Inherited Perils.—Foremost amongst the perils to life, in all its stages, but especially in its early stages, are the inherited. We may safely say that no one is born free from taint of disease, and we may almost say with equal certainty, that there is no definable disease that does not admit of being called hereditary, unless it be accidentally produced. To what is known as specific disease, the disease of diseases; to struma, or scrofula, and its ally, if not the same, tubercular affection; to cancer; to rheumatism and gout; and to alcoholic degeneration, the grand perils of life are mainly due. These are the bases of so many diseases which bear different names; these so modify diseases, which may in themselves be distinct, that, if they were removed, the dangers would be reduced to a minimum. These diseased conditions do not, however, exhaust the list of fatal common inheritances. On many occasions, for several years past, I have observed, and maintained the observation, that some diseases, which are to be noticed in a coming page, as communicable, infectious, or contagious, are also classifiable under this head. I am satisfied that quinsey, diphtheria, scarlet fever, and even what is called drain fever, typhoid, are often of hereditary character. I have known a family in which four members have suffered from diphtheria, a parent having had the same affection, and probably a grandparent. I have known a family in which five members have, at various periods, suffered from typhoid, a parent and a grandparent having been subject to the same disease. I have known a family in which quinsey has been the marked family characteristic for four generations. These persons have been the sufferers from the diseases named, without any obvious contraction of the diseases, and without having any companions in their sufferings. They have been, in fact, predisposed to produce the poisons of the diseases in their own bodies, as the cobra is to produce the poisonous secretion which in its case is a part of its natural organisation.

Accidental Perils.—Next amongst the perils which beset the early life are the accidental dangers to which it is exposed. I do not mean the purely physical accidents, the troubles and blows to which childhood is subjected, not these only, but the more subtle accidents which are incurred through exposure to vicissitudes of season, and to the influence of those particles of the communicable diseases, which, being introduced into the body, incubate there, and transform the secretions of the body into poisons like unto themselves. A long list of diseases incident to the spring-time of life is found in these two classes of causes of diseases, those due to the contagious particles, numbering from twenty-five to thirty alone.

The grand mortality of the child-period is indeed due to the two classes of causes now under our consideration. From exposure to the vicissitudes of season comes, foremost of all, that first step into so wide a universe of evil, the common cold, or catarrh. Upon that comes the continuous visitation which, extending to the pulmonary surface, causes bronchitis, croup, pneumonia, tubercular inflammation; or, extending to the mucous surface of the intestine, causes irritation there, diarrhœa and choleraic affection. From exposure, again, to the poisons of the communicable diseases, there are produced the long and fatal calendars of diseases of shortest incubation, like cholera; of short incubation, like scarlet fever, diphtheria, erysipelas, influenza, whooping-cough, and croup; of medium incubation, like relapsing fever and cow-pox; of long incubation, like small-pox, chicken-pox, measles, German measles, typhus, typhoid, mumps, and malarial fever; and of longest incubation, like hydrophobia. The returns of the Registrar-General will show, weekly, how in persistent procession these diseases march through the land.

Inflicted Perils.—Third amongst the perils incident to the early life are those inflicted by reason of ignorance, or false knowledge and practice, or hard necessity, or all combined. These perils begin with the earliest days of infancy and continue onward. The tight swathing band in which the helpless infant is enrolled, as if it were an Egyptian mummy; the frequent error that is made in depriving it of its natural food—its mother's milk—and in substituting for that true standard of food foods having no proper arrangement nor proper assimilable quality; the too hasty introduction to it of foods in common use in adult life; the not uncommon introduction even of stimulants to these young; the imperfect feeding of the mother, and of pampering her with stimulants when she undertakes the maternal duty of being nurse to her own child; the poisonous method of giving soothing or narcotic 'quieteners' to children; the almost as injurious plan of taking up children from their gentle life-giving sleeps and exposing them to shocks, surprises, and excitements, that are injurious to every function of nutrition and of mental repose; the confinement of the child in close rooms, away from the fresh midday air; the evil plan of taking it out into the night air and into crowds and noisy places, like the railway station or busy thoroughfare; the worse plan still, of scolding, frightening, and even slapping, the helpless being, and thereby implanting in it a nervous, irritable nature, which it will never lose—these are the truly crying evils, which in earliest, dreamiest, and most eventful days and months of human life, plant, imperceptibly, their accursed stings into every day of life that is to follow. If young animals of lower life, that are to be bought and sold and made gross profit upon, were to be subjected to the same penalties, there would be such discomfiture in the selling of them that the reform of the manner would soon be accepted by the most ignorant salesman. It was so in the time of the insane traffic in human flesh and blood. The child of the choice slave, intended ultimately for the market, was often better nurtured in its infancy than the child of the man who owned it, and became a better specimen of humanity.

These evils inflicted on childhood in its first estate are, moreover, followed later on by other evils not less reprehensible, and by one worse than all—I mean the evil of endeavouring, during the time when all the nervous force the growing frame demands is barely sufficient to sustain the natural wants of nutrition, to tax that growing frame beyond the powers that belong to maturity, with competitive mental and physical labours. Both good in their way in moderate form, both necessary for health in moderate form, mental and physical labours are, in these days, made the bane of the nation. The false and useless efforts which crumple up the animal and spiritual natures, making distaste for all labour an early disease, and blighting every flower of genius so soon as it begins to bud, is equal in falsity only with the conviction it engenders, that men and women are made but to learn up to the time of maturity, and that an education which is not what is called 'finished' when the school or college is left behind, is an education that can never be made up in after life. I know nothing so deathly to mind and body as this anxiety—now all but national in its acceptance—to complete education within twenty-one years, when the fact really is that length of life, and length of happy life, depend on the continued cultivation of mental and physical existence beyond all and everything else.

He who has ceased to learn begins to die.

Schools for boys and girls, do you say? 'Yes,' I reply; 'and schools for men and women through every phase of life, if you would have them complete their career.' That crystal brain of the young man, surcharged with more than it can bear, will discharge itself abruptly, and remain an empty centre. But the crystal brain, always crystal, slowly charged and sedately assimilating, will retain its natural lucidity and power through every stage, and will animate to its natural termination the body to which it is the seat of the ministering spirit.

And still to this grand evil inflicted on youth there is a supplemental evil which adds physical to mental scathing—viz. the infliction of corporal punishment on the helpless young before they know why that is wrong for which they are punished, and often when no wise man or woman could detect any wrong leading to the savage performance save the wrong done by the one who punishes. To me, as a physician, nothing is more tainted with iniquitous injury than that corporal punishment of children which proceeds to teach what is believed to be wrong by the instant infliction of physical pain. To the punished and the punisher alike the system is as mischievous as it is barbarous. On the punished it brands hate, or servility, or palpitating fear. On the punisher it brands coward, tyrant, hasty adjudicator of rights and wrongs; while it so perverts the judgment that he who would scorn himself if he struck a woman stronger than himself, will think the act right if a helpless child be the object of his infliction. In another century it will sound as the tones of inquisitorial suffering sound to this, that in our public schools, not masters merely, but masterful boys, should be trained, during the seed-time of health, to tund, to strike with ashen rods, their younger, feebler fellows for faults, or failings, for it may indeed be for virtues, which they themselves are not old enough to comprehend, nor wise enough to rectify.

Acquired Perils.—The perils acquired by the young themselves, acquired as a rule from imitation of the habits of their seniors, form a last part of the dangers incident to this seed-time. In boys, late hours, smoking, resort to the use of stimulants, indulgence in games of chance, and self-infliction of early worry, are special acts ruinous to the foundation of a long and healthy life. In girls, the passion for unhealthy systems of clothing; for compression of the too yielding chest in tight unyielding band and corset; the carelessness about clothing in cold weather; the desire to appear in late evening assembly; the recklessness about food and regularity of meals; the neglect of exercise, and the too frequent fondness of affectation in regard to good common-sense rules of manner and life, are, in their way, as mischievous as the errors committed by the juvenile male community, and in some respects lead more immediately to serious consequences.

We will not, however, dwell longer on this theme, for the faults that might be included in it—were it extended to its full length—would, after all, be found to be but the reflected faults of older humanity: faults irreparable until that older humanity shows the way to those improvements to which it is now necessary to invite your attention.

The New Régime.

I can imagine easily enough that some who are listening to the multiplied evils incident to the seed-time will shrink in despair from all hope of amendment. The sense of necessity of youthful death will seem for a moment to excuse the sense of shame. I hear one, sighing, say: If this be by design, it is vain to meet it. I hear another say: If this be by no design, but by, as it were, an universal accident or fortuitous occurrence, it were hopeless to try to meet it.

For my part, I am beset with no such doubts or fears. If I begin to think of design, the design I think of is poor mine; I am designing for the designer, and must come to grief. If I think of no design, I am merely helping to build up something for those who conjure up design from their own designing. I, therefore, am content to feel assured that, while there is design in regard to this mortal life of man, it is out of the range of my inadequate comprehension; I bow my head and say I do not know. And yet there are lines of thought resting on knowledge of natural facts in which the directions of the design of life are traceable; these are laid, first, in the observation of constantly recurring phenomena bearing on this subject; secondly, in the observation of those phenomena of sentiment or undemonstrated opinion which also bear upon the subject.

Touching, then, the actual recurring phenomena, we may, I think, discover from them most distinctly that the tendency of human life is always towards a more perfect condition; that the natural tendency is towards a more perfected life, and that when man himself does not, in ignorance or intention, do what is injurious to himself, natural law does not. Nature follows truly its own course, and gives us no help against ourselves; but the moment we see the right way she is with us in our efforts, and with giant power helps us on. We are not to natural law as so much inanimate matter; we stand above natural law as we stand above the brutes. As our divine Plato expresses it—'We are plants, not of earth, but of heaven; and from the same source whence the soul first arose, a divine nature, raising aloft our head and root, directs our whole corporeal frame.'

Towards this same view our sentiments converge. We compare all that is desirable to all that is healthy, and the *summum bonum* of our wishes is the *summum bonum* of health. We cling to the idea of a persistent life even beyond death: a life encrowned with such health that to be sick and to die is impossible. We cling to the idea of such a life in unmeasured happiness: a life devoid of pain and sorrow, a perfected health. We cling to the idea of such a life in realms of perpetual beauty: a life of the beautiful of beauties, health in its completed form and character.

Thus, in this instance, reason and sentiment are one, the surest proof of truth.

On the sentiment involved in the proposition I need not dwell: it thrills in every breast. On the reason I am bound to dwell, and if it be but in one instance, I should give proof of it. I will give one; a contrast of good and evil, of health and disease under human direction, and, I may say, under human control.

There were, some years ago, two communities existing at one time, and noted by an able observer. One community was at Montreux, a parish in the Canton of the Vaud, in Switzerland, a parish of two thousand eight hundred and thirty-three souls. The pastor there, M. Bridel, kept a life-history of his charge, and during a long series of years recorded births at the rate of one in forty-five, and deaths of one in sixty-four annually, a death-rate of 15·62 in the thousand. The other community was a Russo-Greek, existing at the same period of time. In this community the births were one in seventeen, the deaths one in twenty-five, or at a rate of forty in the thousand. In the Switzer parish one sixty-fourth died per year; in the Russian, one twenty-fifth, or more than twice as many. In Montreux four-fifths of those born reached twenty years; in the Russian class, six hundred and six out of one thousand perished ere they had attained their fifteenth year, the nuptial garments of the mothers becoming, as it was said, the shrouds of the first-born. In the Swiss community the march of life, seemingly slow, was towards health and an improving life; in the Russian the march of life, seemingly so fruitful, if it had been calculated by the birth-rate alone, was the most fatal in Europe.

I would not, for my part, set up this Swiss parish as perfect—far from it; it was but half perfect. Still, the contrast is before us. Why did it exist? The answer was clear. The Swiss success was due to simple forethought and the virtue of continence. Those civilised peasants of the Vaud conserved their health, their happiness, their life, by the comparative slowness and circumspection with which their successive races were brought upon the scene of the world. Those uncivilised Russian-Greeks, reckless as to birth—not much more reckless than some great English towns have been in our time—lost their health, their happiness, their life, by their mad growth of life. With them death was the shadow of birth, and they had no shame. In our present day, in our best communities, though the reason for the shame is less than it was, yet still in the seed-time of health it is double what it ought to be, or what it need to be. That the reason for it diminishes is proof enough that it may diminish more, until it be refined to the delicacy of susceptibility of those who dared not let the sun behold their young dead.

How towards this perfection shall we wend our course?

We have seen that, in the seed-time of health, there

are four influences at work, sustaining the perils that bring the cause of shame. It is by carefully and earnestly correcting these influences that our course shall be towards success and honourable vitality.

To those *inherited* perils of which I have spoken our minds must first be turned. Say you, the task of reducing them is difficult, delicate? It is all that. But it is not insurmountable in a world that has commenced to throw off its animal impulses, and to reason, and to believe, that 'from the same source whence the soul first arose, a divine nature, raising aloft our head and root, directs our whole corporeal frame.'

I know, and it is hopefullest knowledge, that I shall be listened to by thousands with attention and respect when I urge that, in regard to these inherited perils, wise men and wise women will soon begin to hesitate, even in relation to the marriage tie, before they of a certainty inflict those perils on the world. And with this hesitation such good will come as I dare not express. Let it be known that there are certain marriages which must lead to intermarriages of disease of body or mind; let it be known that results of combinations of this kind are inevitable towards premature death; let it be known that results of combinations of this kind are as inevitable towards sickness and death as combinations of health are inevitable towards health and long life, and we cannot but feel sure that no perversity of folly can long continue to produce through birth the most fatal types of all the fatalities. Let hereditary health be once recognised as an element of the marriage contract, and the health and life of the nation will receive a lease that shall double the value of one and the duration of the other. I speak on this point not from simple enthusiastic hope, but happily from a knowledge singularly cheering. A short chapter of mine in 'Diseases of Modern Life,' entitled 'The Intermarriage of Disease,' has itself during the last six years been the means of checking many of what would have been most deplorable instances of these intermarriages.

While this reform lingers we have some direct means in our hands for lessening the extent of even propagated perils. The tendency of hereditary perils is towards removal when the influences which support them and nurture them are removed. By beginning early in life to place those who are born to peril in conditions for good life, it is astonishing how much can be practically done for them in their bad, if not in their worst, estate. Take as an example of this reforming service the Anerley Schools, where waifs and strays of society, born to all kinds of physical perils, are tended and trained in mental and physical arts. It is like a regeneration. The bloodless, the scrofulous, the rachitic, the rheumatic, predisposed by birth to these afflictions, burst out into such active life that the diatheses seem in abeyance. Nature, always pursuing her unchanging course, would go with a bad system, no doubt, and cure the world of those affected by sweeping them from it, if they were left to their fate. Happily, she goes also with those who work to save, and, aiding them, cures the world by restoring to it its life, and by re-endowing it with health.

In this cause and course the schoolmaster becomes the physician, and the more we have of this branch of the healing faculty the better for us all.

In the removal of the diseases by inheritance there

are, then, two modes of treatment, the preventive and curative: preventive in wisdom of selection of parentage; curative in training those whom no prevention has blessed, into the choicest conditions for health in the seed-time of health.

There is yet another removable cause of these perils which I touch with the lightest finger, but dare not omit. It is indicated on the chart of sin and shame in dark, black, pall-like blot. It is the physical crime which men and women commit when in days of responsible life they acquire to themselves by intemperance and other terrible indulgences those inheritances of crime which pass to their children and proclaim their shame through them. If we could take the world, drowsy in ignorant lusts, and shake it into knowledge here, what crime and shame were saved in one generation none can tell. I know the mass to be reformed is huge as the mightiest mountain, dense as lead. But faith and knowledge in steady action are all-potent even for overcoming this present overwhelming difficulty.

The accidental perils which beset the young in the seed-time of health, and which we accept as evils which sanitarians are bound specially to combat; those serious perils which spring from the exposure of the body to the poisonous particles which produce disease by contagion or infection, come next before us for removal. We call these perils contagious diseases; we call them plagues or pestilences, and, in respect to them, we have learned much that is accurate, and, I fear, much that is inaccurate. What is accurate is, however, the most important. We know the number of these diseases, we know that their number is limited, that it is confined to thirty at the most, and practically to little over half thirty. We know that the members of this class of diseases have different periods of incubation, that is to say, of period intervening between the reception of the poison and the development of the symptoms produced by the poison. We know that the symptoms of the diseases, once developed, run a regular course. We know that some persons are more susceptible to them than others. We know that, to a certain extent, one attack of suffering from many of the diseases is a cause of exemption from a future attack. We know that the diseases assume an epidemic or spreading character, and that each of them has its season, in which its spread is so remarkable that its general course may be charted or outlined as connected with the time of weeks or months or years. And if, regarding the nature of the poisons which produce the diseases, we know least and are most divided, we have, at all events, this precious knowledge, that the poisons themselves are removable and destructible, so that they lie within the range of human control.

What is more, we have the clearest demonstration that while the poisons of these diseases can be generated, cultivated, and disseminated, when the conditions for such generation, cultivation, and dissemination are present, they can also be prevented to such an extent that places which were their favoured homes can be made the places in which they cannot live.

When you enter a court of justice, in some old country assize town, you see to this day lying before my Lord Judge a bunch of rue. My Lord himself may not know what that bunch of rue means, and the man who cuts it and lays it out will give you, if you ask him, the strangest version of the ceremony. 'Some will rue the day when my Lord Judge comes down to try.' That is true, many will rue the day; but the meaning is not there. That bunch of rue was once, not very long ago, the supposed antiseptic or purifier which interposed between my Lord Judge's nose and the fever-stricken prisoners at the bar before him. Once, not very long ago, the gaols from whence those prisoners were brought were the centres of the great pestilent disease, typhus. The men, stived up in those horrid dens, fed with air charged with their own emanations, and fed with food on which they starved, generated the contagion of disease. They were the cobras of society, secreting a poison worse than the cobra's, a poison volatile, subtle, deadly, that would diffuse into the air, and not spare my Lord himself if he came within the sphere of its influence. The gaols then were the foci of fever. But a change took place. Howard, who was as good a sanitarian as he was a philanthropist, and whose rules for the construction of sick hospitals remain model rules to this hour, proclaimed his mission. The gaols began to improve; one improvement of a sanitary kind followed upon another improvement; the results began to arrest attention, and the good that was being done increased and increased with every year. And now, what think you is the triumph? The triumphant result is that in the gaols, the foci once of diseases of the spreading kind and of worst types, spreading diseases cannot practically exist at all. We might lay roses before my Lord to-day instead of rue, or lay the rue on the dock instead of the bench, for the prisoner, in matter of risk from contagion, is actually safer than his judge.

I cannot overstate this lesson. If the homes of those who live in the seed-time of health; if the nursery, the schoolroom, the school dormitory, the playground, were only kept in the same state of physical purity as the model prison, the perils from the accidental diseases, caused by infectious particles of diseases, were soon removed, and the *immortelles* we see on the little graves so thickly laid in cemetery and churchyard were as little called for as the rue on my Lord's daïs.

To you who are interested in the events that occur in the seed-time of health I press this lesson. I press it because of the truth it conveys, the plain, the practical truth, that the simplest means are all that are demanded for the removal of the most fatal of human foes. You are masters and mistresses yourselves of the position. Those shame-faced mourners who would not let the sun see their faults and sorrows were not so much masters of the position, perchance, as you are; had not the dearly bought experience that has been incurred for you. Shall you be less shamed than they when death from accidental causes which you could so largely control comes to your door or enters your domicile? Again I press this lesson, and there is need of it again, for yet another reason. Science, in the main most useful, but sometimes proud, wild, and erratic, is lately proposing a desperate device, founded on an hypothesis clever and specious, but not yet gilded with wisdom or proof, for the prevention of these infectious perils. She proposes to prevent one peril by setting up another. She would inoculate new diseases into our old stock in the anticipation that thereby the new diseases will put out the old. This may be called homœopathy on the grand scale; and if it goes on we may soon see the ranks of sanitarians divided into two ranks, as we see in medicine the regular and the homœopathic practitioners. I pray you be not led away by this new conceit of prevention. In infinitesimals the homœopathic principle may be harmless enough, and on the old adage,

> 'Our doctor is a man of skill,
> If he does you no harm, he will do you no ill—'

it may sometimes seem to compare favourably with heroic methods of cure. But homœopathy on this grand scale, this manufacture of spick-and-span new diseases in our human, bovine, equine, and canine, perhaps feline species, is too much to bear the thought of, when we know that perfect purity of life is all sufficient to remove what exists, without invoking what now is not. I doubt, indeed, whether it were not better to continue in our present imperfect state than venture to make new additions of prophylactic maladies; and content, with Hamlet's sage advice,

> 'Rather bear the ills we have,
> Than fly to others that we know not of.'

By a few rules, in short, which all prudent and wise people may carry out in their own homes, the accidental perils of the seed-time may be kept from the homestead as easily as from the prison-house. Let every man and wife be their own sanitarians and make their house a centre of sanitation. Let in the sun, keep out the damp; separate the house from the earth beneath; connect the house with the air above; once, nay twice, a year hold the Jewish Passover, and allow no leaven of disease to remain in any corner or crevice; let the house cleanse itself of all impurities as they are produced; eat no unclean thing; come back to the first fruits of the earth for food; drink no impure drink; wear no impure clothing; do no impure act; and all the good that science can render you is at your absolute command.

The perils incident to the seed-time of health which I have called *inflicted*, come before us as altogether removable. To remove them skill even is not demanded; nothing is demanded but common human nature and common human sense. That every mother should nurse her own child; that in the early days of life, before the consciousness is naturally developed, the blessed sleep of infancy should be allowed its natural course; that the senses should not be oppressed until they are duly developed; that the quickly breathing lungs should be fed with fresh air; that the yet feeble digestive organs should be supplied with simple food; that the growing body should be clothed in warm and loose garments; these, surely, are practices the simplest people can carry out, practices easier than most which now prevail. Again, that gentleness should be the law of treatment to the young, and that the mind should be taught to know before the body is taught to suffer, that surely is a practice which all can carry out; a practice which both for learner and teacher is easier and better than many which now prevail. Once more, that the growing bodies of our youth of both sexes should be permitted to enjoy the full force of the growing power allotted to them; that such power should be permitted to play its part for their nutrition, so that the body may be endowed with its full maturity; that, surely, is a practice of letting Nature have her free course—in other words, of letting well alone—which all can follow much more easily than most practices that now prevail. Lastly, that the growing mind should be permitted its free and natural course to grow and grow throughout the whole term of its earthly life, and not be killed in its early career by the insane pressure of labours it is utterly unable to bear, or to apply if it could bear them; that, surely, is a practice simplest of all, most natural of all, and most certain for the promotion of intellectual and social advancement.

The fourth series of perils incident to the seed-time of health—those which I have designated the *induced* —are, like the last, entirely under human command. For them to be removed, however, a reform beginning with those who have passed the seed-time is the absolute necessity. These perils must cease, and can only cease, by the process of the younger learning what is right from the examples of the older and wiser creations of humanity. While middle-aged and old men and women indulge in low and injurious luxuries and pleasures, which inevitably shorten and embitter existence; while these revel in intemperance, and break every sanitary law in the Decalogue and out of it, it cannot be expected that imitative youth will do less than follow in their staggering and bewildering footsteps. What now is wanted is the ideal of a new nobility. In the wildboar days of human existence, in the days when men, hardly emancipated from lower forms of life, crept out of their caves, their huts, their walled prisons, to see their nobler species go forth to exercise those rude arts of fighting, hunting, revelling, which formed the whole art of civilisation, there was a nobility which deserved the name, the representative of necessity. But now, when these arts have degenerated into mere childish imitations, mere apedoms of the great past, they are but injurious pretensions for nobility of soul and body. Once noble according to the spirit of their day, they are in this day ignoble. Gamblings and struggles for money, false fame, false hopes, false health, they kill the older, cripple the younger, pervert all. I say nothing but what is good of physical exercise : I would that every school were a gymnasium; I would that every man and woman could ride well, walk well, and skilfully exercise every sense and every limb. I urge only that this example be set, that all exercises, whether of body or mind, be carried out in purest habitude and in accordance with the enlightening progress of the age.

Approaching now the close of my discourse, I find two applications of thought with which briefly to trouble you; one general, the other local and connected with this passing hour. I have tried to bring before you the seed-time of health, the time when this humanity of ours, in body, mind, and spirit, is learning either to live well or to live ill, to live long or to live short, according to its life in the seed-time. I have shown how bad is the seed-time, how pressing the shame of it, and how shameless nevertheless. I have tried to show what are the elements of reform which in that seed-time are required. In general expression of thought I would, respectfully as earnestly, ask those who rule and govern us to look at this period of life as it is; to make it their test object of good or bad government; to assure themselves that when the death-roll of this period of life reports itself filling, filled, the government is bad, happiness out of the question; peace, order, national greatness all impossible; that when the death-roll of this period is emptying, is emptied, all is well; that life then promises to run its completed course, and peace, concord, and prosperity to accompany the health that is ensured.

But to you, Brightonians, I address myself specially. It may easily be your fate, if you will it so to be, to have less cause for shame than even those shrinking mourners of whom I drew a picture in my opening lines. You, planted by the silver sea, have now, in spite of yourselves, a health which you do not of yourselves deserve. You, whose coats the breeze of the sea brushes, whose homes it of its own wild will cleanses, you are made for the work of tending those who are living in the seed-time of health. That specifically, in so far as your resources permit, is your great mission. You have called us sanitarians here to speak the truth that is in us. Let our meeting be useful; let it be the date from which the shame of those mortal events the sun ought never to witness may be felt whenever they occur. You have before you opportunities almost without parallel. You have Nature with you in all her freshness, expanse, and beauty. Learn her ways from herself. Embarrassed by no traditions of antiquarian treasures, you can pull down and rebuild as freely as you can build anew. You are already a school-ground of schools: let that be your abiding tradition, and make your town, in which the ideal of a model city was announced, be the model Hygeiopolis itself, *The Commonhealth of the Commonwealth*. Then your sons, proud of their ancestry, shall realise even here, that ' as is the heavenly such are they also that are heavenly;' and approaching the Infinite Spirit, from whom all proceed and to whom all return, shall declare, not in words merely but in very deeds, that perfected consummation of sanitary principle:—' Thy will be done on Earth as it is in Heaven.'

HEALTH AND RECREATION.

THAT all work and no play makes Jack a dull boy is one of thóse common sayings which we seem bound to accept, whether we like it or not. It is a truthful saying and an untruthful, a wise saying and an unwise, according as one word in it is interpreted, and that word is *play*. If play really means *play* in the strict sense of the term, as it is defined for us in the dictionaries, viz. 'as any exercise or series of exercises intended for pleasure, amusement, or diversion, like blind man's buff;' or as 'sport, gambols, jest, not in earnest'—then truly all work and no play makes Jack a dull boy, and Jill a dull girl.

But in these days there is a difficulty in accepting the saying as true, because the idea of play, especially when it is expressed by the tern 'recreation,' is not always represented in the definition I have given above. We now often really transform play into work; and our minds are so constituted that what is one person's work is another person's play. What a backwoodsman would call his horse-like labour, a foremost statesman may call his light of pleasure. How shall we define it? What is play or recreation?

Men differ, I think, on the definition of work and

play more than on almost any other subject: differ in practice as much as in theory in regard to it. I have had the acquaintance, and I may say the friendship, of a man who lives, it is said, for nothing but recreation, or pleasure, or play. Such a man will rise at ten in the morning, and after a leisurely, gossiping, paper-reading luxurious breakfast will stroll to the stables to look after the horses, of each one of which he is very fond. He delights in horses. Thence he will away to the club, will gossip there, reads the reviews or the latest new novels, and regale at luncheon. After luncheon he will play a rubber, winning or losing several shillings—it may be pounds. He may then take a ride, or drive, or walk in the Park, and have a chat there; or canter over to Kew and look round the gardens; or attend a drum; or visit the Zoological or Botanical Gardens. After this he will return home, and, ably and artistically assisted, will dress for dinner. The dinner, in accordance with his life, will be elegant, sumptuous, entertaining, whether he take it at his own table or abroad. After dinner he may probably go to a ball and dance until two or three in the morning; or if there be no ball on hand, he may have another rubber, or a round at billiards, or a turn at the play, the Opera, or the concert-room, with a final friendly chat and smoke before retiring for rest.

To this gentleman—and I am pencilling a true and honest gentleman, not a modern rake of any school of rakes—this mode of life is a persistent pleasure, and to many more it would, I doubt not, be a perpetual holiday. To me it would be something worse than death. The monotony of it would be a positive misery, and I am conscious that many would be found to share with me in the same dislike.

Some will say that is all true enough with respect to persons who have passed out of youth into manhood, but that when life is young the distinctive appreciations for different modes of recreative pleasures are not so well marked out. I doubt, for my own part, this belief. It seems to me that in childhood the tastes for recreative enjoyment are as varied as they are in later years, with this difference, that they are not so effectively expressed. The little mind is ever in fear of the greater, and is often forced to express a gladness or pleasure which it does not truly feel. When children, left to themselves, are independently observed, nothing can be more striking to the observer than the difference of taste that is expressed in respect to the games at which they shall play. More than half the noise and quarrel of the nursery is, in fact, made up of this difference of feeling as to the character of the game that shall be constituted a pastime. In the end, on the rule, I suppose, of the survival of the fittest, the strongest children have their way, and one or two little tyrants drag the rest into their own delights.

I should, on the grounds here stated, venture, then, to say that there is, in point of fact, no more actual difference between work and recreation than what exists as a mere matter of sentiment: that recreation is a question of sentiment altogether, both in the young and the old.

If we could get this fact into our minds in our educational schemes for the young we should accomplish at once a positive revolution in the training of the young, which revolution would, I think, be attended by the happiest change and train of thought in those who, in the future, shall pass through the first stages of life to adolescence and maturity. The search for amusements, and for new amusements, amongst the well-to-do would not be needed, since the mind from the first would be naturally brought to find a new delight in each act now called labour. The word 'labour,' in short, might drop altogether; the praise of labour, which is so often extolled, would find its true meaning; and the blame of play, which is so often unduly criticised, would have its proper recognition.

It has always seemed to me that in that once high though brief development of human existence; in that period, if we can believe that the art of the period came from the life of it when the human form took its most magnificent model for the artist still to copy; in that period when the perfection of bodily feature and build indicated, of itself, how splendid must have been the health of the living organisations that stood forth to be copied and re-copied for ever—it has always seemed to me, I repeat, that in that wonderful period of Greek history, so effulgent and so short, the reason why such physical excellence was attained rested on the circumstance that amongst the favoured cultivated few, for they were few after all, there was from the beginning to the end of life no such thing as work and play. Everything was existence—nothing less and nothing more. Every office, every duty, every act must have been an existence for the moment, varied but never divisible into one of two conditions, practical pain or practical pleasure. Life was an enjoyment which nothing sullied except death, and which was purified even from death by the quick consuming fire, that the life might begin again instantaneously and incorruptibly.

If by some grand transformation we could in our day approach to this conception which has been rendered to us by the history of art, and could act upon it, we should, in a generation or two, attain a degree of health which no sanitary provision, in the common meaning of that term, can ever supply. If we could turn our houses into models of sanitary perfection; if we could release our toiling millions from half their daily labour; if we could tell want to depart altogether; if we could give means of education to every living human being; we should not remove care, and therefore we should not secure health unless with it all we could also remove the idea of the distinction of labour and pleasure, the morbid notion that some must work and some must play, that the world may make its round.

In this country, so differently placed to the country of the great and the ancient nation of which I have spoken, it is impossible, perhaps, ever to introduce a joyousness like to that which the favoured old civilisation enjoyed. Our climate is of itself a sufficient obstacle to such a realisation. Where the physical conditions of life are so unequal, where we waste in structure of body, whether we will it or not, at certain fixed seasons, and gain, whether we will it or not, at other fixed seasons, it is impossible to attain such excellence by any diversion of mind or variation of pursuit. For universal gladness the sun must play his part, doing his spiriting gently, but never actually hiding the brightness of his face. From us, for long intervals, his face is hidden. Under these variations of the external light and scenery around us we have to cripple our minds through our bodies. Our clothing must be heavy during long stages of the year, and our food so comparatively heavy and gross, that half the power which might otherwise go off in vivacity, or nerve, or spirit is expended in the physico-chemical labour that is demanded for keeping the body warm and moving and alive.

To these drawbacks is added the unequal struggle for existence; the partitioning off of our people into great classes, millions who are obliged to work from morning to night, and of thousands who are at liberty to make some change in their course of life: the millions who are tied to some continuous, monotonous round of labour, until the whole body lends itself to the task with an automatic regularity which the mind follows in unhappy and fretful train, with little hope for any future whatever on earth that shall bring relief: the thousands who would be better off if they had more to do.

From whatever side we look upon this picture it seems at first sight to present an almost insoluble problem, when the conception of mixing recreation with work, so as to make all work recreative, is considered. Amongst the masses there is no true recreation whatever, no real variation from the daily unceasing and all but hopeless toil; nay, when we ascend from the industrial and purely muscular workers to the majority who live by work, we find little that is more hopeful. There is no true recreation amongst any class except one, and that a limited and happy few, who find in mental labour of a varied and congenial kind the diversity of work which constitutes the truly re-creative and re-created life of man.

We get, in fact, a little light on the nature of healthful recreation as we let our minds rest on this one and almost exceptional class of men of varied life and action of a mental kind. They come before us showing what recreation can effect through the mere act of varying the labour. The brain-worker who is divested of worry is at once the happiest and the healthiest of mankind, happiest, perchance, because healthiest; a man constantly recreated, and therefore of longest life.

The late Dr. Beard, of New York, computed the facts bearing on this particular point, and left us a reading upon it which is singularly appropriate to the topic now under consideration. He has reckoned up the life-value of five hundred men of greatest mental activity: poets, philosophers, men of science, inventors, politicians, musicians, actors, and orators, and he has found the average duration of their lives to be sixty-four years. He has compared this average with the average duration of the life of the masses, and he has found that duration to be fifty years in all classes the members of which have survived to twenty years of age. He, therefore, gives to the varied brain-workers a value of life of fourteen years above the average. By a later calculation, relating to a hundred men belonging, we may say, to our own time, he has discovered a still greater value of life in those who practise mental labour, seventy years being the mean value of life in them. Thereupon he has enquired into the cause of these differences, so strange and so startling, and has detected, through this analysis, as I and others have, a combination of saving causes, the one cause most influencing being the recreative character of the work. His observation is so sound, so eloquent, and, above all, so practical, I can feel no necessity for apology in giving it at length. He is comparing, in the passage to be quoted, what he calls the happy brain-worker with the mere muscle-worker; and this is the argument :—

'Brain-work is the highest of all antidotes to worry; and the brain-working classes are, therefore, less distressed about many things, less apprehensive of indefinite evil, and less disposed to magnify minute trials, than those who live by the labour of the hands. To the happy brain-worker life is a long vacation; while the muscle-worker often finds no joy in his daily toil, and very little in the intervals. Scientists, physicians, lawyers, clergymen, orators, statesmen, literati, and merchants, when successful, are happy in their work without reference to the reward; and continue to work in their special callings long after the necessity has ceased. Where is the hod-carrier who finds joy in going up and down a ladder; and from the foundation of the globe until now, how many have been known to persist in ditch-digging, or sewer-laying, or in any mechanical or manual calling whatsoever, after the attainment of independence? Good fortune gives good health. Nearly all the money in the world is in the hands of brain-workers; to many, in moderate amounts, it is essential to life, and in large and comfortable amounts it favours long life. Longevity is the daughter of competency. Of the many elements that make up happiness, mental organisation, physical, health, fancy, friends, and money—the last is, for the average man, greater than any other, except the first. Loss of money costs more lives than the loss of friends, for it is easier to find a friend than a fortune.'

The contrast put before us in these forcible remarks is most striking. It is the key to the position in trying to unlock the secret as to what true recreation should be. These brain-workers of whom Dr. Beard speaks are, indeed, the modern Greeks, not perhaps in perfection, but in approximation. The Greeks might possibly have gone higher than they did in the way of developed physical beauty and of mental endowment, and these happy brain-workers of later ages might perhaps still more nearly approach the Greeks. But both were on the lines towards the highest that may be attainable, and this, as a means of indicating the right line, is my reason for using the illustrations that have been offered.

That which I have so far urged consists, then, of two arguments. Firstly, that recreation to be healthful must, as its meaning conveys, literally, be a process of re-creating; that is, of reconstructing, or rebuilding; a practice entirely distinct from what is called play, when by that is meant either cessation from every kind of creation, or enjoyment of abnormal pleasures which weary mind and body. Secondly, that they who are able to live and recreate in the manner suggested are, in positive fact, they who present the healthiest, the happiest, and the longest lives.

From these premises I draw the further conclusion that we have no open course of a reasonable kind before us except to strive to beget a healthful recreation in the direction indicated.

At the same time I do not say this in order to divert attention from what may be rightly called the natural animal instincts of man. I have no doubt there might be a cultivation of mind which should cease to be recreative, and which thereby should be as injurious to the health of the body as an over-cultivation of mere gross mechanical labour, and which might even be more dangerous. It is not a little interesting to observe that the greatest of the Greeks had become conscious of this very danger, as if he had learned its existence from observations in his daily life. Plato, in treating of this subject in one of his admirable discourses, warns us against the delusion that the cultivation of nothing but what is intellectually the best is, of necessity, always the best. It is more just, he says, to take account of good things than of evil. Everything good is beautiful; yet the beautiful is not without measure. An animal destined to be beautiful must possess symmetry. Of symmetries we understand those which are small, but are ignorant of the greatest. And, indeed, no symmetry is of more importance with respect to health and disease, virtue and vice, than that of the soul towards the body. When a weaker and inferior form is the vehicle of a strong and in every way mighty soul, or the contrary; and when these, soul and body, enter into compact union, then the animal is not wholly beautiful, for it is without symmetry. Just as a body which has immoderately long legs, or any other superfluity of parts that hinder its symmetry, becomes base, in the participation of labour suffers many afflictions, and, through suffering an aggregation of accidents, becomes the cause to itself of many ills, so the compound essence of body and soul, which we call the animal, when the soul is stronger than the body and prevails over it—then the soul, agitating the whole body, charges it with diseases, and by ardent pursuit causes it to waste away. On the contrary, when a body that is large or superior to the soul is joined with a small and weaker intellect, the motions of the more powerful, prevailing and enlarging what is their own, but making the reflective part of the soul deaf, indocile, and oblivious, it induces the greatest of all diseases, ignorance. As a practical corollary to these remarks, Plato adds that there is one safety for both the conditions he has specified : neither to move the soul without the body, nor the body without the soul. The mathematician, therefore, or any one else who ardently devotes himself to any intellectual pursuit, should at the same time engage the body in gymnastic exercises ; while the man who is careful in forming the body should at the same time unite the motions of the soul, in the exercise of music and philosophy, if he intends to be one who may justly be called beautiful and at the same time 'right good.'

Such is the Platonic reading of the recreative life as it appeared to him in his day and amongst his marvellous people. We have but to trouble ourselves with half the problem he refers to, and with but half the advice he suggests. Little fear, I think, is there amongst us that the soul should be so much stronger than the body, and so greatly prevail over it that it should agitate the whole inwardly, and by ardent application to learned pursuits cause the body to waste away. Nor is this to be regretted because if the danger so stated were a prevailing one we should have two evils to cure in lieu of one which is all-sufficient for the reforming work of many of the coming generations of men.

I have not, I trust, dwelt too long on what I may call the practical definition of recreation as it ought, I think, to be understood, as it was once understood and practised, and as it is still practised, if not systematically understood, by a few whose varied and delightful works and tastes make them the healthiest and longest lived amongst us.

It is well always to have a standard before us, though it be seemingly unapproachable, and the illustrations I have endeavoured to supply of all work and all play, and of long-continued recreation thereupon, form the standard I now wish to set up for observation.

To make all England—and all the world, for the matter of that—a recreation-ground: to make all life a grand recreation: to make all life thereby healthier, happier, and longer: this is the question before us.

Confining our observations to our own people and time, it may now be worth a few moments of analytical inquiry as to how far we, in different classes of our English community, are away from so desirable a consummation—the consummation of all human effort towards the perfected human life: the dream of some poets that such a life has been and will return—'Redeunt Saturnia regna'—the dream of many poets that it is to be, if it has not been.

The Registrar-General, with much judgment—due to long and wide experience of the component parts of the nation comprised under the title of England and Wales —has divided our community into six great classes, which classes are in many respects so distinct that they may almost be considered as great nations of themselves, having their own individual pursuits, habits, tastes, and, if the word be allowable, recreations. He described from the census of 1871. (1) A professional class, of governing, defending, and learned persons, numbering some 684,102 persons, chiefly of the male sex. (2) A domestic class, wives and women of the household, and hotel and lodging-house keepers—a large class, the great majority women, numbering as many as 5,905,171—nearly, in fact, six millions. (3) A commercial class of buyers, sellers, lenders, and transporters of goods and produce, chiefly men, and numbering 815,424. (4) An agricultural class, cultivators, growers, and animal keepers, the majority men, numbering 1,657,138. (5) An industrial class, mechanics, fabric manufacturers, food and drink producers, and purveyors of animal, vegetable, and mineral produce—a very large class, having in it members of both sexes, and numbering 5,137,725. (6) An indefinite non-productive class; persons of rank and property, and scholars and children, with nearly an equality of representation of numbers of both sexes; the whole class including a total of 8,512,706, of whom 7,541,508 are scholars and children—the living capital of the next generation of men and of women.

As we glance at these classes we quickly detect that what may be called their vocations are extremely different; that each class—with the exception, perhaps, of two, the professional and the commercial, with that part of the indefinite class which is composed of persons of rank and property, and which approach each other—are as widely separated in tastes and habits and inclinations as they are in labours and works. Looking at the education of body and mind in these classes as a whole, there is certainly little enough of symmetry.

Amongst the representatives of these classes which are best able to command the advantages of true recreation there is little sound attempt to use the privilege in a refined and reasonable way. The persons who have their time at command, and who belong to the most favoured division, are divisible into two groups. (*a*) A group which does no work at all that can bear the name of useful or applied labour, but which spends all its waking hours at what it considers to be recreative pursuits, which may be laborious, but must not be remunerative. (*b*) A group which labours industriously for the sake of return or reward, but which steals from time of labour regular intervals in order to follow out certain of the recreations which form the whole life of the first group, in strict imitation of that envied group, and in hopeless neglect of any recreation of its own better adapted to its real wants and best enjoyments.

Each of these groups suffers from the course it follows. The representatives of the first kind lose much, since they are for ever repeating the same, to them, pleasurable or automatic activity. The second lose because, while they are ever repeating the same useful activity, they are only relieving that activity by repeating day after day the same automatic and imitative recreations. Thus both are subjected to what may be called the automatism of recreation, an automatism of recreation bad in every sense, and specially bad in the present day, because of the quality of it, as well as the limited quantity, and because there is in it no such diversity of recreation as is wanted to keep the body in health by the exercise of the mind. With one man the recreation is all taken out in cards, with another in chess, with a third in billiards, with a fourth in debate or gossip on some one persistent topic of discourse or argument, and so on, for what may be called the indoor recreative life. Nor is it much different with outdoor recreative amusement. Some one particular amusement claims the attention of particular men, and to this amusement the men adhere as if they had to live by it, and as if, in fact, there were no other recreative pursuits in the world.

This speciality of recreative pleasure or labour—for soon it becomes labour—leads to consequences which are often of the most serious character. The man who undertakes the recreation at first as an enjoyment, and, indeed, as a relaxation, becomes so absorbed in it that he strains every nerve to be eminent in it, a professor of the accomplishment, with a local repute for his excellence. The moment he enters on this resolve, he loses recreation. He sets himself to a new work, be it mental or physical; his mind becomes a nursery for the produce of that one particular culture, and he is in respect to that not far removed from a monomaniac. From the day that he is completely enamoured of the special pursuit it is little indeed that he is good for out of it in hours apart from the common vocation of his life. He becomes fretful if for a day he be deprived of his peculiar gratification; irritable if he joins with others in it who are not so skilful as himself; envious if he meets with a rival who is better at it than himself; and often actually sleepless in thinking and brooding over some event or events that have been connected with the previous play or venture.

If the time at my disposal admitted the introduction of detailed illustration of the facts here referred to I could supply from experience instance upon instance. I have seen an amateur chessplayer so infatuated with the game, which he originally sat down to as a relaxation, that he became for months a victim of insomnia. He carried the whole chessboard, set out in various difficult problems, in his brain, if I may use such a simile, studied moves on going to sleep, dreamt of them, woke with the solution solved, was sick and feeble and irritable all next day, followed his usual occupation with languid ability and interest, resumed his play at night with excited but not recruited determination, got more and more sleepless, and at last failed to sleep altogether. I have known more than one similar illustration in whist-players and in great billiard-players, and have seen the results of these so-called recreations end in the most sad physical disaster, when the pursuit of them has been made a matter of living importance, and when the player has ever had in his mind that pitiful *if*: ' If I had done this or that—if I had made that move on the board—if I had played that card—if I had made that stroke, how would the case have been?' It matters little what the answer to the question may be—whether it be that by such a move, card, or stroke the game would have been lost or won; the perplexing doubt is there to annoy, and it keeps up an irritation which imperceptibly wears out the animal powers and does permanent injury to life. You see men while still they are actually young grow rapidly like old men under this supposed recreative strain. They grow prematurely careworn, prematurely grey, prematurely fixed in idea and obstinate in idea, angry at trifles, baffled by trifles, and, in a word, young senilities.

In every busy city there are hundreds—may I not extend the calculation and say thousands?—of men who, in pursuit of the recreative pleasures I have specified, or of others similar in their results, are wearing themselves out twice as fast as they need do by the legitimate labour to which they have to apply themselves in order that they may earn their daily bread. The observant physician, as he listens quietly to the statements of these suffering men, is obliged, in his own mind, to differentiate between the assigned and what is often the real cause of that train of evils to which it is his duty to lend an attentive ear.

Thus, amongst the most intelligent part of the community—amongst the part that can help itself—there is no systematised scale or class of recreations that can be relied upon to afford the change really demanded for health. Nor are matters much improved when we take up the kind of change that is sought after by the same classes in the matter of physical recreation. When the Volunteer movement first came under notice, and for some time after it first came into practice, it was the hope of all sanitary men,—I believe without any exception,—that the exercise, and drill, and training, and excitement which would be produced by the movement would prove most beneficial to the health of the male part of the people at a period of life when the training of the physical powers is most required and often most neglected. I remember being quite enthusiastic at that change and its promises, and I recalled the other day an often-quoted paper or essay which had sprung out of that enthusiasm, and which I dare say at the time it was written seemed common-sense itself. I can but feel now that the hope was begotten of inexperience. The movement has been a success, I presume, in a national and political point of view, but a careful observation of it from its first until this time has failed to indicate to me, as a physician, that it has led to any decided improvement in the health generally of those who have been most concerned in carrying it out by becoming its representatives. Certain it is that nothing affirmative of good stands forth in its favour, and I wish I could stop with that one neutral statement. I cannot in order of truth and fairness so stop, for I have seen much injury from the process. To say nothing of the expense to which it subjects many struggling men, to the loss of time it inflicts on them, to the neglect it inflicts at the fireside and home, to the spirit of contest of mind and fever of mind which it engenders; to say nothing, I repeat, of these things—all of which, nevertheless, are detrimental, indirectly, to the health of the men themselves and of those who surround them in family union —there is a direct harm often inflicted by the service, call it recreation if you like, which is not to its credit.

The man who has advanced just far enough in life to have completed his development of growth, and to have lost the elasticity of youth, the man who has rather too early in life become fat and, as he or his friends say, puffy, the man who has, from long confinement in the office or study, found himself dejected and dyspeptic, each one of these men has passed into the ranks of the Volunteers, in order to regain the elastic tread, to throw off the burthen of fat, or to find relief from the dyspeptic despondency. For my part, I have never been able to discover a good practical result in any of these trials; but I have seen many bad practical results. I have seen the partly disabled men, in the conditions specified, striving to do their best to keep alive and be on a level with younger and athletic men, and I have been obliged to hear of the signal and natural failure of the effort. I have heard of the attempts to meet the failure by the tempting offer and too willing acceptance of what are called artificial stimulants to give temporary support, and I have been obliged to discover in persons so overtaxed and so overstimulated a certain heavy excessive draw on the bank of life, an anticipation of income which, in the vital as surely as in the commercial world, is the road to a premature failure and closure of the whole concern.

There are many who will agree with me, I doubt not, on this point; there are many men, and there are more women,—for wives and mothers are far more observant and wise than husbands and fathers on these points,— who will be able to bring their experience to bear in confirmation of that which I have spoken; and these will agree that to put men of different ages and of different states of constitution and habits in the same position for recreation; to trot them all through the same paces; to make them all wear the same dress, walk or march the same speed, carry the same load, labour the same time, move the limbs at the same rate; that to construct one great living machine out of a number of such differently built machines is of necessity an unnatural and, in the end, a ruinous process. There are some, however, who, while admitting so much, will put in a plea for the young members of the community. They will insist that the younger men, men who are from nineteen or twenty up to twenty-nine or thirty, may, with advantage, go through the recreation of training after the Volunteer fashion. The case is much stronger on behalf of this argument, but even in the respect named there requires a great deal of discrimination. A race of strong men may be bred, and a weak race may, by gradual development, be raised into a strong; but a weak man, born weak, can, through himself, be led a very little way into strength; while during the process of training he can most easily be broken into utter feebleness, so that the last of the man may be worse than the first. Hence, in training the weak into strong through any form of recreation, mental or physical, but specially physical, there must be a singular discrimination. In this instance of Volunteering as a mode of progress in physical health for the young there are dangers that ought to be avoided with religious care. To advise a weakly youth of consumptive tendency and feeble build, or one having some special proclivity to rheumatic fever, heart disease, or other well-defined hereditary malady, to compete with other men of the same age and of athletic nature, in the same recreative exercise, is to deceive the youth into danger. To force such a one into violent competitive exercise, and tax him to the same degree of vital withdrawal day after day, or week after week, is to subject him all but certainly to severe, if not fatal, bodily injury.

I have selected the recreative exercise of Volunteering as a case for illustration of an important lesson, and I have made the selection, not because the recreation is special as a sometimes harmful recreation, but because more persons are concerned in it just now than in aught else of the same kind of recreative pursuit.

There are many other so-called recreations which are even more injurious to the feeble adolescent and to the enfeebled matured individuals, who seek to find symmetry of health in extreme recreation. Football is one of these recreations fraught with danger. Rowing is another exercise of the same class. Polo, while the fever for it lasted, was found to be of similar cast. Excessive running and prolonged and violent walking,— in imitation of those poor madmen whose vanity trains them to give up sleep and all the natural ordinances that they may walk so many hundred miles in so many hundred hours,—these are alike injurious as physical recreations, unless taken with the same discrimination as is required by those who enter into the Volunteer movement.

As we pass from the freer and wealthier classes of the community into the less prosperous we find no marked improvement whatever in any form of recreation. We begin, in fact, to lose sight of the recreation that ministers to either mind or body in a sensible and healthy degree, and to see that which should be recreative replaced almost entirely by continuous and monotonous labour. The idea of symmetry of function and development between mind and body disappears nearly altogether; so that, indeed, to mention such a thing would, in some of the classes concerned, be but to treat on a subject unknown, and therefore, as it would seem to them, absurdly unpractical. To tell a country yokel that his body is not symmetrical in build, and that his mind has no kind of symmetrical relation to his body, were cruel, from its apparent satire. Yet why should it be? Why should ignorance and labour so deform any one that the hope of a complete reformation, the hope of the constitution of a perfect body and in it a perfect mind, should seem absurd? It is not the labour that is at fault. The labour is wholesome, healthful, splendid; it is a labour compatible with the noblest, nay, the most refined of human acquirements. Why should it be incompatible with perfect physical conformation of mind and body? It is not, indeed, the labour that is at fault, but the ignorant system on which it is carried out.

There is much difference, in fact, between the three classes of the community called the domestic, the agricultural, and the industrial, in respect to the work, the recreation, and the resultant health pertaining to each class. The domestic class as a whole is, by comparison with the industrial, fairly favoured. The members of it lead, it is true, a monotonous life, and see often but little of the beauties of external nature, but they find in the amusements they provide for those who are about them some intervals of change which are, as far as they go, of service. Moreover, except in that part of the class which is engaged in disposing of spirituous drinks, and which pays a heavy vital taxation from the recreation springing out of that vocation, its representatives are not exposed to harmful recreations to any extreme degree. The domestic class therefore presents, on the whole, a fairly healthy life. The majority of its members are women and mothers; and, in the gladness with which they tender their love and adoration to the young and innocent life that comes into their charge, they find, perchance, after all, the purest pleasure, the most entrancing, the most ennobling recreation, that, even in the midst of many cares and sorrows and bereavements, falls to the lot of any section of the great community.

The agricultural class, less favoured in recreative

opportunities than the others which have passed before us, living a laborious and very poor life, ever at work for small returns, and finding little recreation beyond that which is of mere animal enjoyment, is still comparatively favoured. To the agricultural worker the seasons supply, imperceptibly, some delight that is beneficial to the mind.

> ' These as they change, Almighty Father! these
> Are but the varied God.
> Mysterious round! what skill, what force divine
> 'Deep felt in these appear: a simple strain,
> Yet so delightful, mix't with such kind art,
> Such beauty and beneficence combin'd,
> And all so forming one harmonious whole—
> Shade unperceived, so soft'ning into shade
> That as they still succeed they ravish still.'

The labour of the outdoor agricultural class, blessed by these changing scenes which the exquisite poet, above quoted, so exquisitely describes, is varied also in itself. Each season brings its new duty: the spring its meadow-laying and sheep-shearing; the summer its haymaking; the autumn its harvesting and harvest-home, and fruit-gathering; the winter its ploughing and garnering, and cattle-tending; with sundry well-remembered holidays which are religiously kept. There may be through all this continuous, wearing labour; there is; but, as it is not monotonous, it is to some extent recreative, and the facts of mortality tell that it is saving to life. The agricultural classes present a mortality below the average in the proportion of ninety-one to one hundred of the mass of the working community. Moreover, there is hope for the agricultural classes in the fact that it is, comparatively, an easy task to supply them with a perfect roundelay of beautiful recreations for their resting hours. It is only to remove from them the grand temptations to vice in the beershop and the spirit-store, and to substitute for these resorts a rational system of enjoyments, to win for the country swain the first place in that symmetry which Plato called 'right good.'

The utter blankness, the blankness that may be felt, in respect to recreation is realised most in the millions of the industrial class who live in the everlasting din of the same mechanical life; who see ever before them the same four walls, the same tools, the same tasks; who hear the same sounds, smell the same odours, touch the same things, feel the same impressions, again and again and again, until the existence is made up of them, never to be varied until death doth them part. It is to this class,—repining, naturally envious, naturally restless, and at this moment of time unsettled, mournful, and disaffected, to an extent which few, I fear, of our rulers comprehend,—it is to this class most of all that the balm of wholesome recreation is most necessary, and for whom the absence of it is most dangerous. In this class there is no such thing as health. It is a blessing not to be found. You could not, I solemnly believe, bring me one of them that I dare, as a conscientious physician, declare, after searching examination, to be physically healthy in any approach to a degree of standard excellence. As a rule the average of life amongst those who have passed twenty-five would not be above fifteen years.

In these classes we see the effect of what I may venture to call the denseness of work, leading to mortality in the most perfect and distinctive form; work without any true recreative relief; work without anything changing or becoming recreative in itself; work relieved at no regular intervals for introduction of new life.

The greatest of all the *social* problems of our day is involved in this study of the manners and modes of thought of over five millions of adult English people, all confined in order that they may labour, with no satisfactory relief from labour, and with no land of promise before them. The greatest of all the *political* questions of our day is also involved in this same study. The physician knows that the wisest of mankind, the most intelligent of mankind, are only half their former selves when they are out of health. He knows that health which is bad, but not sufficiently bad to prostrate the physical powers to such an extent as to cause inactivity of the will, is the most perplexing of all states of mind and action with which he has to deal. He feels thereupon a fellow-sympathy with the political physician who is called upon to treat the industrial masses in mass; to provide for their minds' health, to calm their excitement, to plant confidence in their hearts, and—most arduous task of all—to find out the way for securing for them those two grand remedies in the Pharmacopœia of the ordinary physician, rest and change of scene, in pure and open air.

'They find their own recreations, these working millions,' I think I hear some one say. They *try* to find them, would be the truer statement. They try their best, but they have found few conducive to health, many that are fatal. They are to be pitied and pardoned for these errors of their finding. What if they do discover recreation of the worst kind in the bar and saloon of the spirit-seller ? Have they not the example of the wealthier classes before them, teaching that the same indulgence, in another style, is recreation ? May they not ask how many other obtainable pleasures are provided for them, and whether many—too many—of obtainable pleasures so called, and so bad, are not positively thrust upon them? They have laboured all day in monotony : where shall they go for recreation and what shall the recreation be ? If they go far away, they are removed from the sphere of their labours ; if they look near to their own abodes, they find not one true and ennobling pastime, but fifty that are degrading, and, at the same time, filled with every possible temptation.

I apply this to our own people; but it is, I fear, equally applicable to other peoples. Dr. Beard, the American I have already quoted, wrote his experience, gathered in his own country, as follows : 'To live,' he says, speaking of the same classes, 'to live on the slippery path that lies between extreme poverty on the one side and the gulf of starvation on the other; to take continual thought of to-morrow, without any good result of such thought; to feel each anxious hour that the dreary treadmill by which we secure the means of sustenance for a hungry household may, without warning, be closed by any number of forces, over which one has no control; to double and triple all the horrors of want and pain by anticipation and rumination—such is the life of the muscle-working classes of modern civilised society; and when we add to this the cankering annoyance that arises from the envying of the fortunate brainworker, who lives at ease before his eyes, we marvel not that he dies young, but rather that he lives at all.'

There remains still in the list of classes requiring recreation, and the health that springs from it, the last or indefinite class. Of the poor in the indefinite class I need not speak; for they—the waifs and strays of our civilisation—are, I fear, under little influence of such refining agencies as we would put forward for the future. With the very small class of persons of rank and property, less than 169,000 altogether, I have dealt already, by joining them with the professional and commercial well-to-do classes. To the seven and a half millions of scholars and children and their recreations attention will be called in a new chapter.

HEALTH AND RECREATION FOR THE YOUNG.

THE study of the recreations of those who are in their early years brings up one of the most pressing questions of modern social life. There is so much diversity of opinion and of practice on this subject that no kind of system can be said to prevail in relation to it in any class of society except the poorest, in which necessity rather than choice enforces a gloomy and sad uniformity. If there be any rule at all it is, perhaps, that parents and those who have charge of the young give to them those kinds of recreation which they themselves were taught to enjoy. There is much haphazard about the matter at the best; and, may I be allowed to say, there is much that is called recreation which has no reference to health, and which therefore is not recreative, although it pretends to be.

On the other hand, many recreations which may be healthful to the young become hurtful, owing to the times and modes of carrying them out, and from no other causes. The holding of children's parties at late hours in winter time is one of the most dangerous examples of this kind, as it is one also of the most foolish of modern devices. The children's party is now often called for seven or even eight o'clock at night, and young children, for three or four hours after they ought to have been in bed and fast asleep, are kept up in the midst of a feverish excitement, which will not cool down for two or three succeeding days. In this excitement they are frequently fed with foods and drinks of the most indigestible character; and from the excitement, the dyspepsia which results, and the colds which are engendered from the exposure to cold air late at night and when the strength is exhausted, there is set up, almost of necessity, temporary derangement of the body, and in some cases fatal disease.

There is a fact to be ever borne in mind in respect to exposure of the body during these inclement seasons; and the fact is this: at these seasons the body is undergoing a natural process of waste or consumption—a veritable loss of weight, which is a cause of exhaustion, and which is increased by every additional exposure. This waste is in progress during all the winter months, from November to April, and affects the whole of the community. Thus the danger increases as the winter months progress, until it attains its maximum in the fatal months of early spring, when so many of the enfeebled of all ages pass away.

There is another and indirect danger connected with children's parties which I must here incidentally notice, although it lies a little apart from my subject, because of its immense and practical importance. In the cold seasons, when the body is at its lowest working power, the epidemic diseases are often most rife, and are always most dangerous. These diseases are also at such times communicated more easily from one person to another, the poison which produces them being brought into a company, not in the open space, but in the close room. Thus the winter children's party becomes not unfrequently the centre from which an epidemic takes fresh root, and is, in fact, a focus of spreading disease. There is much thoughtlessness on the part of grown-up people, who have little people in their charge, in this particular. I knew, lately, an instance in which some young children who were recovering from scarlet fever, and who could not leave their rooms, were allowed, as a pastime, to make the dolls' clothes and the decorations for a Christmas tree, which dolls' clothes and decorations were to pass a few hours later into the hands of a large party of juveniles, to the certain infection of some, who in going there to play would be going possibly to death.

Dr. Whitmore, a late and able medical officer of health for Marylebone, most forcibly called the attention of the public to this source of danger, and illustrated the fact of danger from the clearest evidences of mortality. But, in spite of all, the imprudence still goes on, and I must be pardoned for one minute of irrelevancy in having once more referred to it.

To return to our subject proper. The next series of dangers in the recreations of the young are contracted by over-competition and by the equality of effort enforced alike on children and youths of different build and constitution. Nothing can be more absurd, nothing, indeed, more cruel, than the inflictions which, in the name of recreations, are perpetrated in the manner named. Let me give one illustration.

I was at a swimming bath, where some twenty boys,

all under twelve years of age, were swimming or learning to swim. There was no comparison between these lads in matter of physical outline. One, a short lad, had a narrow and pointed chest, a fragile form, an almost transparent skin, a chilly surface of body, and a blue lip; another, a tall, broad-shouldered, broad-chested youth, had a full and ruddy complexion, a warm surface, and a firm muscular build. The others varied between these two standards. Such were the physical conditions of the swimmers, and so constructed, physically, they commenced their recreation. Swimming is a healthful recreation, when properly carried out, as well as useful and necessary. But here was the mischief from it in the case of the boys in question: there was no discrimination in the amount of the recreation. The boy with the pigeon chest and blue lip had quite as much of it as was safe for him at the end of five minutes; but, to keep him up to the same standard as the swimmer with the broad chest, who had such buoyant lungs that he could hardly sink even if he tried, the weak boy was encouraged and driven to go on and keep on until he had passed through the same exertion as his more favoured comrade. The result was that the weak boy came out of the water blue all over, an hour later was as blue as a bilberry in the lips and cheeks, and was cold, shivering, feeble, and sleepy. I could see those boys as plainly as if I had followed them going back to school and to the afternoon work. I could be as sure as I could be of any physiological fact that it would require six hours, under the most favourable of circumstances in relation to food, rest, and warmth, to fully revitalise that feeble boy up to his own imperfect standard, while no number of hours would ever bring his vitality up to the standard of his more fortunate fellow-student.

Look now at the error, at the long series of errors, committed by this mode of recreation on the feeble boy. His animal warmth had been robbed unduly, and he was therefore languid and unhappy. His blood was aerated less freely than it should be, and he was therefore circulating blood more slowly than he ought, and breathing with excess of labour. He was more susceptible to every depressing influence, and his nervous system was dulled in the same manner as it would be from sitting in a close and badly ventilated room. He would be drowsy, and perhaps the master of his school would say idle or apathetic. For this he would be rated at his lessons, compared with other boys who got on better, and, if not punished corporeally, made irritable and anxious in mind, which is another form of punishment. This nervous lad, never over-strong, would be again unduly taxed. And now, what else must follow as results? When the nervous system is low and depressed, the digestive power is enfeebled. When the digestive power is enfeebled, the nutrition of the body is degraded in every part. Then the vital organs, on which life depends, and in which the activity of nutritive changes ought to be most rapid and regular, are the first to suffer, while even such passive parts as those which make up the skeleton do not escape scot-free. See, then, what a modification of healthy life may be easily effected by one apparently trifling error of recreation. Let that error be repeated many times, or let some equivalent error be performed and repeated many times, and what is the almost necessary evil? The almost necessary evil is the institution in that feeble body, in active form, of the phenomena of disease towards which it had a proclivity, and on which the feebleness depended.

It will possibly be urged by some that the process of making these feeble boys compete with stronger boys is intended to invigorate the feeble. For a similar reason, these same naturally enfeebled children are often sent out of doors in cold weather to '*harden* them.' The ignorance is beyond pardon. As you cannot gather grapes of thorns nor figs of thistles, so you cannot out of a weak animal frame extract strength except by taking it out of the bank of life, to the premature shortening of the inherited store of life—a store which, *ceteris paribus*, may be fairly calculated from the mean value of life in the two latest generations of the stock from whence it has been derived.[1]

---

[1] The calculation may be cast as follows after an actual calculation in which the estimated value of the last life proved correct within a year:—

Actual value of paternal grandfather's life . .	80 years.
Actual value of paternal grandmother' life . .	64 ,,
Actual value of maternal grandfather's life . .	68 ,,
Actual value of maternal grandmother's life . .	72 ,,
Actual value of father's life	71 ,,
Actual value of mother's life	71 ,,
Total value of the six lives	426 ,,
Estimated value of last life 426÷6 . . .	71 ,,

Actual value of last life, 72 years less one month.

In these calculations, deaths from accidents and from accidental diseases have to be excluded, and much labour, therefore, is required for collecting as many facts as would form a basis for a positive rule. I have collected sufficient to indicate that, allowing a range of five years

By care in training we may make this life extend to its full term, or a little over, but only by careful conservation. There is no making up for what is once positively lost in the matter of life. Life is the reflex of the dying earth in this respect. 'There are three things,' said the Caliph Omar, 'which come not back—the sped arrow, the revengeful thought, and the spoken word.' He would have added a fourth had he been a vital physicist, viz. the stroke of the heart.

It is of no use opposing these natural facts; we might as well buffet rocks with our hands. In training up the child towards his natural standard we must in everything thoughtfully conserve; we may use up power to its bearable limit—for that is exercise in the true sense, and is necessary—but to do more is to destroy. To get a stronger, and longer-lived, and finer model of human kind, we must change through progenies, not through individuals. We must alter the factors, then the figures will come.

Mothers know these facts better than fathers, and mothers are often laughed at because their knowledge is the knowledge of love, and passion, and anxiety rather than of cold, calculating, reasonable—I had almost said commercial—expectation. But the mothers are none the less right, and, indeed, men know better when they are dealing with lives that have to be bought and sold. The great breeders of flocks and herds do not try to harden or over-tax the young lower animal they want to perfect.

---

of estimated value on either side, the method above stated affords a fair basis for a general rule of calculation of the hereditary value of human life in this stage of our knowledge of the means for preserving and maintaining life.

They use a wise discretion, and they succeed in what they do. At the same time they are often indiscriminate about their own children. A country surgeon, whom I much esteemed for his quick insight, once brought to me for consultation a feeble boy of a consumptive tendency, in order to settle the question whether the mother's fostering or the father's hardening system should prevail. 'The father,' said my friend, 'is a clever man; he is most successful in the management of cattle, and if he would be only half as clever in the rearing of his children all would be well. But he is very hasty on this point, and the other day—a day bitterly cold—he did two of the most inconsistent things I ever knew a man to do. He quarrelled like a fury with his poor wife for sending Charlie to skate with his legs in warm stockings, and five minutes later dismissed his groom for taking a colt out for exercise without clothing it in a horse-cloth.'

The instance I have adduced of swimming is all I can afford time for in illustration of excessive physical recreation. The example is one of fifty or more in which errors of a similar kind are perpetrated in the management of the young. Yet is there one more I dare not let pass by, because of its great importance. It relates to the plan of forcing recreation on children by surprises or by mere force or insistence, against their courage rather than against their will.

A child during some recreative exercise is told to do something he would do if he could, but which he dare not do. He is told to go into the water; he is told to jump over a fence or a wide gap; he is told to mount a horse or to get on the bar at the gymnasium. The child hesitates, and then, too often, comes the risk of danger in direction. The child hesitates, and thereupon he is admonished or he is bidden not to wait, or he is even made to do what he is bidden; or, worst of all, he is tricked into the act, under the impression that once he gets through the ordeal he will care for it no more, hesitate never again.

This is disastrous work, beyond any measure of comparison. Little seeds of evil are sown in the mind by these proceedings, which grow into the most terrible consequences. Distrusts are engendered, and doubts as to the good faith of the nearest and dearest, and therefore of every one. So there is developed a distrustful mind, which is of all minds the most pitied for its own sake and for everybody's sake that has to do with it.

And when we come to analysis of facts in relation to causes, we find again that the practice which leads to these sad distrusts is as foolish as it is hurtful. Courage is not a quality that can be infused into a child by threat, and trick, and force; it is a quality men and women are born with, and its centre is the heart, not the head. No one can make a person with a physically feeble heart courageous. We say of men or children who are strong and courageous, they are lion-hearted; we say of those who are opposite, they are chicken-hearted; and the terms express the facts. But as we cannot by the most consummate skill transform a chicken into a lion, so we cannot make a chicken-heart a lion-heart. We can encourage, set example, explain the freedom from danger, explain how to do the thing that looks dangerous without much exposure to risk; and so we can train even a faint-heart to become morally, if not physically, brave. But to try to give physical courage to a body that is weak at its centre; to try to force courage out of such a body; to try to call forth what is not there; to make it an opprobrium to be weak-hearted; and, under the name of coward, to hound shamefully a poor, fluttering, gentle, loving, trustful nature, as is commonly done, is one of the wickedest pieces of ignorance with which I am conversant. It never makes a man brave, but it makes many assume bravery, and by the means of assumption generates a race of cowardly hypocrites who are the very curses of social life.

I have touched on one or two of the most deadly errors connected with the recreative exercises of children. I might pass now to the consideration of those modes of recreation which promise to be most conducive to national as well as to individual health. In this task I might cast back on the different classes in the reverse manner to that I have hitherto followed, beginning with the younger and proceeding to the older members of the community.

Before, however, I enter into details of this character, I would like to clear the way by referring to certain often styled recreative pursuits or pleasures, which, as I think, ought to be removed altogether from that position.

There are four classes of so-called recreations which deserve to be placed under the condemnation I have named. The exercises included in these different classes may be called pastimes, or games, or accomplishments, or amusements, or anything else. My argument is not with them in that sense; my argument is that they are not recreative, and therefore are not healthy.

Firstly, then, whatever calls forth the passion of expectation for the sake of self-interest is not recreation but destruction. I mean by this, that whatever so individualises a human being that in the pursuit of it all thought concentrates in himself and his own selfish expectations and hopes for success, exposes that human being to a risk greater, perchance, than the risk he is speculating upon. The risk means anxiety or worry; the anxiety means a sense of fulness and oppression within the chest, and that sense of oppression means an undue pressure and load upon the heart. In course of time, sometimes prolonged, at other times instantaneous, the motion of the heart, under the excitement, loses its nervous balance, and then there is set up a truly physical condition of broken heart—a condition in which the heart intermits in its movements, or beats out of rhythm, without respect to time. This is, in fact, a broken heart—a heart no longer steady, no longer ready to meet emergency or carry its owner comfortably into the vale of a long life.

The exclusion of such influences on life as those to which I now refer throws out of the order of recreation all games in which what are called *stakes* are played; or it would be more correct to say, all resort to games in which the game is used for the purpose of play for high stakes. The games may in themselves be innocent enough, and even recreative, when they are merely intended for simple exercise of skill. In this sense a game at whist, or other games at cards, or a game at billiards, bagatelle, chess, or the like may be purely recreative and useful; but when stakes come into the play in such manner as to excite great anxiety and expectation, then comes the danger. Winning then and losing then is in either case bad. Winning elates; losing depresses; both destroy.

I know of little that has been a worse physical scourge to the human race, in civilised life, than this system of using recreation for the purpose of winning or losing, or, as it is called, staking. I say nothing of the moral injuries at all; they are not in my province. I speak of the physical; they are in my province. And of this I am certain: that no man, woman, or child can indulge long in any chance game for more than trivial stakes, and remain in health. Health and chance are incompatibles. Whoever tries the experiment has ceased to find recreation, and may say most truly with the worst of adventurers:—

> For I have set my life upon a cast,
> And I will stand the hazard of the die.

To the young, whatever partakes of a tendency to indulgence in games of chance should never be taught, and the desire for it should ever be suppressed or diverted by some more wholesome recreation. To the older members of the community the temptation to the same presumed recreative pleasure should equally be withheld by all wise and prudent monitors.

It is a good sign of our times that the taste for games of chance is steadily passing away, and I think our Government never did a wiser, a more healthful, a more national, a more rational act, than when with firm and unhesitating voice it forbade the mad project to establish a public lottery for the presumed relief of the sufferers from the Glasgow Bank failure. There were doubtless many broken hearts—hearts physically broken —from that sad calamity; but the number was a bagatelle compared with the number that might have been reckoned, had that appalling scheme for the promotion of national degradation been allowed to run its ruinous course and to set up its ruinous example.

Secondly, whatever calls forth a craving or fixed and overpowering desire for the repetition of any particular pleasure or gratification is not a recreation, but a destruction. In craving, as in gambling, individuality overcomes the better judgment. The organic or vegetative part of the nature of man conquers the reasoning, and a self-possession is attained, which in its extreme form leads to the maddest of crimes—robbery, forgery, falsehood, lust, suicide, murder. Whoever craves unduly for anything is, strictly speaking, unbalanced in mind, and is practically insane. Whatever, therefore, ministers to the animal part of man so entrancingly that it leads to intense desire for repetition, or craving, is not recreation, but destruction.

The growing intensity of craving for a pleasure is the most solemn danger connected with it, and, what is worse, it is in most instances a danger which, once trifled with, is long endured. The wisest cannot escape it, nor the strongest, nor the best, when it is once established, without an almost mortal conflict. One of the wisest, one of the strongest of minds, one of the best, the great Sir Humphry Davy, discovered, in his scientific researches, a singular fascination, which with him passed into what some would call a recreation, in the breathing of laughing gas. In this process he became absorbed and lost in such luxurious dreams that all the universe seemed to him to consist of nothing but thoughts. He revelled in dreams that at times reached ecstasy. At last the craving for this false, this factitious existence became to him so extreme that he could not watch a person breathing, could not look at a gasholder, without experiencing the intense desire to be once more indulging in his aërial nectar. This was an exceptional delight, to which he alone became accustomed, but it is only typical of many that are more common and equally dangerous, which, acquired in early life, are the after-penalties of some part of life, and which, acquired even late in life, are not free from their evil consequences.

Such kinds of so-called pleasures are not recreations; they are destructions: the body is not recreated upon them nor by them. They kill time, and time is life; and so they kill life, for they shorten its days. Drinking strong drinks; taking into the body narcotics and narcotic fumes, as the smoke of tobacco; eating too much of the assumed good, but really bad things of this life; these are the luxuries which beget the fatal cravings that are most injurious. I will not venture to offend by putting in too strong terms the denunciation of such forms of pretended recreative pleasures. I will let the matured who recreate after this fashion remain as the scapegoats of the immatured, and I will simply enforce that the cravings to which I have specially alluded, and all others that might be referred to, ought not to be cultivated between the period of birth and the attainment of the majority of any child born in this era of civilisation.

Thirdly, whatever in the way of a pleasure or delight shortens the hours of natural repose is not a recreation, but a destruction. If it break repose outright by the circumstance that it keeps its victim out of the way of going to rest at proper hours, it is not recreative; if it allows its victim to go to bed at proper times, but keeps him awake in thought, and restlessly striving for sleep, it is not recreative; if it permits him so far to sleep as to let him lose his consciousness of external things, but forces him to dream, it is not recreative. Almost all recreations, as they are believed to be, which introduce strife, or competition, or chance, produce this effect. They are not recreations at all. They do not recreate; they destroy.

One illustration of this form of injury occurs to me here as very practical, and as important because it relates to a comparatively new and increasing danger. I refer to an exercise that is day by day becoming more and more popular amongst our young and untrained population, that of indulging in fiery and systematic debate on the most solemn and abstruse questions. It does not matter in these debates whether or not a debater should show carefully studied knowledge; the point to be gained is to secure a victorious contest, to win at all hazards. It is not even necessary that the debater be consistent in the course that he takes, for it may be that his reputation rests on the fact that he can, with equal skill, discuss the subject successfully on the opposing sides of it. No; what has to be cultivated is perception, finesse, the trick of catching up from an opponent some point on which to found an adverse argument; to throw at one time fire and at another time water on the heads of opponents. These are the seductive arts which govern the young debater, and which set up cross vibrations in the fibres of his as yet growing and unformed brain.

For my part, I think there is far too much of debating amongst the educated classes in all periods of life. I never see a man of culture, who possesses the qualities necessary to become a teacher, enter the arena for mere showy debate, without a pang at seeing what energy is thrown away, that might be expended on thousands of ignorant outsiders whom to teach would be a national blessing. The sight presents to me the picture of a number of well-fed citizens going into a ring and throwing their rich viands at each other's heads and immensely disfiguring themselves, while a vast multitude outside is howling for the mere necessaries of existence.

But when we come to the young debaters, then we come to the crisis. Then is the time to see the pelting heart, the flushed brain, the straining expression after what is not known, the heated declaration often of what the speaker would afterwards give anything to withdraw, the fierceness of expectation, the flush of conquest, the pallor of defeat, the babble of discord, the succeeding restlessness, the weariness without repose, and the resolves and schemes for the future :—then, I say, is the time to see these things and to consider what they will bring forth.

This is not recreation, but destruction. If I dared to lift the professional veil and show the mental havoc which I, as one only, have witnessed from this form of contest, you, my readers, who are not learnedly conversant with the facts, would wonder little at my earnestness. It is the saddest part of this subject that those men or youths whose minds are most excitable, least reasoning, most impulsive, least absorbent, are the youths who are most given to wish for the contest, and are most liable to suffer physically and mentally from its results. If they fight through the early ordeal without injury they are fortunate, and even then are not benefited; for when they are young they acquire a debating, controversial disposition, when they are old they do not depart from it, and, according to their relative power, they bore to its very vitals the comfort of a family, a town, a corporation, or it may be a nation, until they create a wholesale rebellion against themselves, in which at last their very friends join, and they are left to the inevitable fate of being easily beaten by cooler and keener, though perchance less endowed intellectual opponents. To conclude this head, nothing is recreative that does not naturally lead to repose. That is a simple rule to remember, and still simpler to act upon. When any one feels, by a few observations, that anything he does, be it ever so pleasant, interferes with his natural repose, let him be assured that, whatever pig he may have caught by the ear, that pig's name is neither Health nor Recreation.

Fourthly, whatever is rendered automatic in mind or act is not recreative and is not conducive to health. Automatism long continued becomes, in fact, a form of slavery, makes the mind fretful until the automatic process is carried out in due time and order, and thereby makes both body and mind fidgety, so that rest is not obtained in a regular and systematic way. For aged persons automatic amusements are, it is true, less harmful than for the young, but I have no doubt that even the aged are far more benefited by the pleasures of changing recreations than by any orderly and systematic recurrence of one particular pleasure. Variation prevents undue pressure and wear on a single centre or set of nervous centres; it also prevents the sense of periodical restlessness until something to be done is done, and so it conserves the life.

Some nominal recreations of a *physical* kind are, under all conditions, so extreme that they ought to be tabooed by all sensible and civilised people at all periods of life. At the top of the list of these bad physical exercises I place football. This game, in some modes of playing it, is the cause of more physical mischief than I can describe. To say nothing of the immediate injuries that occur from it by falls, sprains, kicks and concussions, broken bones, dislocations, broken shins, and other visible accidents, there are others of a less obvious kind, which are sometimes still more disastrous. Hernia, or rupture, is one of these disasters; varicose veins is another; and disease of the heart from pure over-strain is a third. One of the finest built youths I have ever seen, who came directly under my own observation, was for two years entirely disabled owing to the excessive action of the heart induced by his becoming a champion at football; and he escaped well to recover at all. The weak do not recover perfectly at any time.

This is not recreation, but destruction; and how it is that in the present day of enlightenment there can be found masters of schools who encourage the worst forms of such a savage, damaging, right-down insane pastime is one of the wonders of the day.

Middle-aged men and men past that period do not, as a rule, play at football, but lately they have taken, in vacation time, to a recreation which is to them almost as bad, and that is climbing. At home a man may find the second flight of stairs up to bed as much as he cares to do in the way of ascending; but in autumn, after ten or eleven months' hard work, he thinks he must invigorate himself by climbing a mountain that has become celebrated for its difficulty. He thereupon buys an 'alpenstock'—I think that is the right word—and up the mountain he goes as far as he can blow. Perhaps he does what he wished to do and gets down again, and then he wonders why he is worse for the effort! why his breathing is so embarrassed, and why he should feel so much older; while his friends wonder that he, who was climbing Swiss mountains a few months ago, should have died so suddenly—such a healthy-looking man, so active, and only, after all, in his prime. His friends would not wonder if they knew the strain which he, already inelastic and incapable of strain, had passed through in his great effort.

Sir Walter Raleigh is said to have etched on a window-pane—

Fain would I climb, but fear lest I should fall;

and his queen is said to have added—

If thy heart fail thee, do not climb at all.

It would not be bad practice for every middle-aged man who is ambitious to climb a mountain before he dies to ask his physician whether, if he climbed at all, his heart would fail him or let him down low beyond recall.

Let me pass from these considerations to those which relate to the amusements that are demanded in our present modes of life. I sometimes see by the public comments on my papers, an inference is being drawn that I am unfavourable to all recreation. Nothing could be more incorrect. True recreation is, in my opinion, one of the grand necessities for health during every stage of rational life.

In my previous essay I referred to the fact that amusements such as chess, which call for mental effort, cease to be recreative so soon as they degenerate into hard mental labour and leave impressions firmly fixed on the mind. I named chess because I believe that of all games it calls forth the largest share of mental labour, and that it easily ceases to become a recreation. At the same time, I have nothing but good to say of it when it is resorted to occasionally without imposing upon the player any great tension or mental strain. I think it is, in moderate taste of it, a very fine mental exercise, which mixes well both with physical and mental work, and which is thus, in the purest sense of the term, recreative. And so of other similar mental sports which involve no chance or stake.

Nay, in games of chance themselves, such as cards, I see nothing but good recreation when the stake, which is their sting, is extracted from them. Whist, as a study of proportion in numbers, is a fine mental exercise—a mental kaleidoscope full of pleasant surprise and wonder—while those games which depend partly upon chance and partly upon skill, such as billiards, combine in their legitimate application a mixture of mental and physical exercise, which is excellently recreative and healthy when healthily pursued. The only objection to these games that can be raised against them is their easy degeneration from recreation and health into labour, and worry, and weariness, and disease.

It is part of my experience, indeed, that whenever any one cannot indulge in recreation, and whenever any one is so busy with work that recreation seems to be a bore; or is so depressed and dull that recreation becomes a penalty—then any man, woman, youth, or child, is in danger. It is, in all such examples, ten chances to one that the person so circumstanced is suffering from some physical malady, which is in turn affecting and enfeebling the mental powers.

My objections are all directed against false notions of recreation, against prodigal expenditure of time and labour in assumed pleasures or pastimes which wear out the body and mind instead of recreating the one and refreshing the other.

I am led then to ask, What are true and natural recreations? What is absolutely necessary in the way of recreation for persons of different ages, different stations, and different modes of life?

Let us first consider the subject in relation to the early periods of life.

In England now the whole of the youth of the kingdom is now under educational control. The institution of universal education a few years ago marks an epoch in the national history. Magna Charta was not a nobler page. In the Board schools the minds of the young of all the masses are turned towards what is good or bad, and as in them the example of the high-class voluntary schools will probably be largely followed, we may fairly assume that we have, through the young, the recreation as well as the learning of the nation fairly in hand, and with this realisation of power we should assume that recreation ought, in fact, to become a real part of the educational programme.

We ought, in other words, to make the subject of

recreation a scientific study, so that natural recreative delights might be put on the proper line for serving health.

Viewing the subject in this sense of it, I should place music as the primitive of recreative pleasures. It proves itself first by its spontaneity. We mark that our children are well and happy when they can sing. We see men and women gathered together, and find the height of mirth and happiness when somebody gives a song or a tune. In the most refined society music is the joy of life; in the lowest dens, men hardly above animals, when they meet to be amused, sing. It may be that in all these positions the music is very bad, but it is there, and it extends through creation.

>Hark! hark! the lark at heaven's gate sings.

In a word, this music is an element of nature. It fills the universe; it fills the microcosm of the universe—the human soul.

Here, therefore, is the first recreation to be scientifically studied. Make a nation a musical nation, and think how we have harmonised it socially, morally, healthfully. We cannot begin to teach this recreation too early or too soundly.

We ought to begin by making the learning of notes in succession—the scale of musical sounds—coincident with the learning of the alphabet. The one could be taught just as easily as the other, and would be retained as readily, perhaps more agreeably. Next, the intervals should be taught in a simple but careful way, so that melody may be acquired and the art of sight-singing laid. From this elementary basis should follow the simplest forms of time, after which a plain melody could be read with as much ease as the reading of the first story-book. Simple part songs, leading to endless delight, would succeed in exercise, and a true and natural language in sweet sounds would be the property, in one generation, of all the nation.

The system of teaching the very young useful information from the study of natural objects, in the midst of recreative enjoyments—not in fixed attitudes on hard seats for hours at a time—is another blending of recreation and work which would tend in the most excellent manner to that equalisation of work and play which would be the *summum bonum* of happiness. The *Kindergarten* is an admirable blending of this kind, and is worthy, in our English life, of general imitation. I am myself no slavish admirer of German acquirements; I do not believe that every *Fräulein* who pretends to teach music is a great musician, or better than those who are to the manner born. I do not believe that every German professor is a philosopher, or that every miserable secondhand English pedant who can play no other part than that of a German translator should be allowed to pipe down all native talent as it exists now, and as it has existed in such masters as Faraday, Davy, Locke, Hume, Bacon, Shakespeare, and the hardly mortal Newton. But it is nevertheless true that in mode of education the Germans can teach us many things, and in the *Kindergarten* they have set a lesson which we may with much advantage learn and practise.

Together with these recreations I should place dancing as another recreation for the young; by which I mean not mere set dancing after the form of high-class, high-company quadrille alone, but good, graceful dancing in figures as varied as the changing sky. There are a number of good old English dances which deserve to be reintroduced for this purpose of recreation. A clever schoolmaster could write a school book on dancing that would be a fortune to himself and a source of happiness to all who practically studied it.

After dancing I should put forward for the young of both sexes the process of drill and gymnastic exercises. Swimming, too, as a recreation should, with due care and encouragement, be taught to both sexes. Swimming has a double, nay, a treble, purpose : it teaches a very useful and necessary accomplishment; it is a good exercise, expanding the chest and giving play to the limbs; and it encourages cleanliness, for a good swimmer learns to like the sensation of a clean surface of the body. To that veteran sanitarian Mr. Edwin Chadwick, whose introduction of the half-time system into factory life places him amongst the great liberators from practical slavery, this last advantage of swimming would probably seem on the whole the most healthful.

For recreation out-of-doors, for boys and girls, there is nothing finer in winter than skating, nor for boys has there ever been invented a finer summer exercise than English cricket. The founder of the Olympian games deserves a monument not half as high as the inventor of that immortal game with the ball, and the bat, and the wicket. Cricket has lost in gracefulness, and I, for one, think, in skill, of late years, by the introduction of the somewhat animal exhibition of swift and round bowling ; but it is recovering from that insanity, and I hope it will retain the first place amongst the outdoor games of English boys and men wherever they may go.

There are some gentler games which are good for both sexes out-of-doors. Since its introduction, the game of croquet has been of immense benefit to the health of girls and women. It has taken them out of the house, and encouraged them to activity in pure atmospheres. Badminton and lawn tennis have the same good influences.

There is another game which ought to belong to women as well as men, which is also singularly good, and which should be reintroduced into every village in England. I mean the old English game of bowls. There is no game that calls forth better or more healthful exercise. It calls for skill, it brings every muscle into play, and it does not suddenly exhaust by single and violent paroxysms of effort. It encourages repose, and, as Martin Luther thought, it helps digestion.

If we could teach our young to sing harmoniously, and know the language of sweet sounds; if we could teach them the music of motion in the dance; if we could make them float and move gracefully in the water; and if we could let them recreate in such gambols as I have named, bringing the members of both sexes as often as possible together in innocent and recreative enjoyment —we should indeed give health; we should indeed make a new people, born to health and all its blessedness. Why should we not? We have all the means at our command. We want only the will.

We can deal with the young easily now, if we like, in respect to recreations, and can mould them as we please. But what of those industrial and agricultural adults the millions of workers whose minds are formed, and who wait, and waste, and strive, and still wait?

I put forward the Greek model of recreative life in a former paper as perfect, in its way, for producing an ephemeral type of perfected physical form and beauty. I have been reminded over and over again, since then, that this perfected people nevertheless fell; fell an easy prey to barbarous encroachment; fell, says one of my learned critics, like ripe apples from a fruitful tree.

It is true, and they would fall again under the same conditions. Yet their fate none the less forcibly illustrates my argument. They proved at least what could be done by a section of a great community. Their fate proved no failure in the matter of accomplishment, but a failure in the foundation on which the accomplishment was laid; and that same fate might easily be ours. That is the fate we have to avoid, and the avoidance of it consists in making the causes of the attainments of the favoured minority extend, in some fair proportion, if not in perfection, to the whole of our population. If we would live by perfected knowledge, we must, if I may so say, pin every man to the earth by it. Then there will be a foundation, solid, and unshifting, and satisfied.

Here, therefore, is the problem. How shall we diffuse recreative pleasures amongst the masses?

I have been a great deal amongst these masses. Two public inquiries which I have had in hand relating to the health of these masses have led me into their own centres of life, and to diagnose, with that knowledge which comes from a life of experience in diagnosis, their physical condition; and the fact I have learned—I am not speaking on matter of opinion at all, but of fact—is that the first step to take will be to reduce their hours of labour. I am quite sure that by such reduction they would do more work, and that soon they would do happily in eight hours what they languidly do now in ten. This effected, the next step is to improve their recreative opportunities, by clearing away the loathsome temptations which beset them in every step of their course, so as to make sin an expensive difficulty; and by introducing such pleasures as are harmless and truly recreative.

All those recreations to which I have referred above come in to our aid here, with others fitter for men. The music class? Yes. The dancing class? Yes. The swimming bath? Yes. Cricket, bowls, tennis, drill, the gymnasium? Yes. And to these I would add still other sources of enjoyment; museums for them to enter whenever they have time; free galleries of artistic beauty; an improved stage; a very flood of good and wholesome literature; and colleges in which subjects of advanced knowledge and thought may not be debated, but taught to them *secundum artem*.

I do not think that these improvements are out of the range of accomplishment, any more than are the means for the advanced recreative education of the young. We have all the appliances. It is will alone that is needed.

The six millions of the domestic class of this country —the women of the household—must not be left behind. Their fate follows the fate of the rest, or marches with it. They are married to their fate, and the tie is a close one. But I need not specially discuss it; it is included in what has been, and in what remains to be, said.

We come at last to the recreations of that minority —less than two millions in all—which forms the governing and commanding mental force of the country, the headship of the whole. In this minority—including in it the commercial as well as the professional classes—ought to be found the nearest approach to the perfection of recreative enjoyment, and of all the health that springs from such enjoyment. There cannot be a doubt that the health of this minority is higher than that of the majority, and that its mortality is relatively lower. This is, however, due rather to protection from direct depressing causes, such as actual want, privation, worry, and care, than to affirmative good arising from judicious methods of recreative pursuit. If, indeed, we survey the whole field of recreation amongst these more favoured classes, the inference to be obtained is that their relaxations and pleasures are, on the whole, detrimental to health. The pleasures include, amongst the most prominent evils, late hours; too free indulgence in rich and indigestible foods; indulgence in stimulants and narcotics; a great deal of chance play; attention to many so-called artistic delights which are neither chastening nor ennobling in character; participation in feats of mere animal strength—few of dexterity, and none which are specially invigorating either to soul or body; the perusal of a literature which is not of the highest class; and the encouragement of a drama which, abused most unjustly for its sensational commonness, is never systematically supported in the sustainment of its nobler purposes, as set forth by its grandest representative, 'to show Virtue her own feature, Scorn her own image, and the very age and body of the time its form and pressure.'

We look round in despair, in short, to find a recreation for the favoured few that is healthy in itself, or that, being healthy, is conducted healthily. But the favoured classes are they that should set the example to the rest of the community; and to them it is most urgent to appeal, that they may introduce such reforms as shall be the examples for the many.

The examples are not different from those suggested for the younger and the poorer sections of the community, though they may be supplemented by others which the less wealthy could not afford, to the great advantage of trade and commercial activity. Riding or walking leisurely through the whole of our beautiful country, until all its lovely scenery is appreciated, and the people of different parts of it are known—is one of the recreations for both sexes of the wealthy which would advance their own health, increase their knowledge, and encourage a most useful and fair expenditure of the good things with which they are blessed beyond their fellow-countrymen. And, when their own country is exhausted, there is still all the world before them.

HEALTH THROUGH EDUCATION.[1]

IN this address I propose to consider the question of 'Health through Education,' that is to say, the study of those methods of education by which the mind, during the whole period of its work, may be maintained in a healthy and properly balanced condition, its powers usefully employed, and its natural tendencies allowed full and natural scope and development.

Up to the present time the progress of science for the promotion of health has had reference, almost exclusively, to the physical health in education, to the state of the schoolroom, to the diet of the scholar, to the clothing, to the training and exercise of the body, to the position of the scholar at the desk, and to such-like purely physical considerations. These considerations can scarcely be over-estimated. I have had the happiness to be associated with the most earnest and energetic of the sanitary leaders who, in our generation, have striven to force them on the attention of a public not always too willing to listen to them, and I regret that I should have to put them somewhat aside for the present hour. But I feel there is another subject even of more pressing

---

[1] Address delivered at the Conference on Education, held in the Rooms of the Society of Arts, January 16, 1880.

moment, and therefore I turn to it. The purely physical study has made its way to some extent: the subject I have now before me has made, practically, no way at all, although its importance can hardly be exaggerated.

Men engaged steadily and systematically in taking different views of the same object are led to see differently and to express themselves differently. I cannot therefore conceal that I approach the argument I would set forth with a perfect knowledge of the fact that I must speak what is, or what may seem to be, contrary to the opinions which are entertained by many who are deeply interested in the work of education, and who, in most respects, are masters or mistresses of the argument on its practical, scholastic side. Those who are engaged in the actual labour of teaching from day to day may entertain views very different in kind from mine. Those who are anxious and over-anxious for the education of their children may entertain views of a very different character from mine, and may, indeed, be far more likely than the teachers of their children to differ from me. The teachers will, I think, in their hearts, be in most respects with me altogether.

When I say that the physical side of the health question is not a part of my present programme, I do not quite state the whole truth, for the physical side of the question is, in one direction, admitted in it. There is always in progress a reaction of the mind on the body which, when it is clearly understood, is seen to be momentous in its results. The amount of physical disease that is dependent on mental influence is large beyond any accepted present conception of it. I am almost afraid to express what I know on this point, lest I should appear to be putting forward what is speculative instead of what is real. And yet I may venture to say that of the deaths of adults who die in their prime from what are called natural diseases a good fourth is due to diseased conditions of body that have been induced by mental influences. The actual and immediate cause of the demise, the killing blow, may be outside the body, may be independent of the body, may be very subtle and seemingly very slight, may admit of no correct scientific exposition at this present stage of science, may be some unknown or obscure meteorological influence; and yet the conditions leading up to the point when slight causes take effect may all the while have been in steady progress, and may all the while have been mental, mental from the first, in the persons affected. Thus men in the prime of life often die suddenly from some slight external influence of a physical nature which has acted upon them fatally, and which gets the whole of the blame; but the conditions of body which have rendered that external influence effective have been long in operation; have been, in the strictest form of expression, mental influences modifying the physical structures, and making those structures susceptible of destructive change from slight external shocks or vibrations. Thus, again, hereditary tendencies, originally formed from mental action, are often transmitted in the character of hereditary physical disease, under which, from some slight external influence, death may occur.

Impressions traversing the senses into the organ of the mind afford the most striking illustrations of physical derangements and of degenerations from mental action in which the mental and the physical most intimately blend. They give rise, in fact, to a term which is as distinctly physical as any that would describe a mechanical concussion or blow—the term, most correct in its application, of 'mental shock:' a shock or blow received by the body through the mind, and producing physical action in the body; a transmutation of an unknown force, which we have only named, so far, by metaphysical names, such as fear, anger, hate, love, into a strictly physical force and a resultant effect; a vibration through the senses, yet not of mere sound, not of mere light, but of something more of which hearing or sight are but the modes of conveyance, modes of conveyance into the nervous atmosphere or ether, to be changed there into some new state of motion or into a new physical condition that is inimical to continuance of life.

Let me explain by one example.

A little boy was once brought to me by a medical friend under the following painful circumstances. The boy was the son of a carpenter, and his father sent him occasionally to a neighbouring timber-yard to give orders for wood. The keeper of the timber-yard, a modified type of Mr. Quilp, had a morbid delight in frightening children. He had bought a large ugly and savage dog, and he tied the dog closely up in a recess in the passage leading to the timber-yard. The little boy I speak of, knowing nothing of this new and terrible importation, was proceeding, as usual, down to the yard, when the dog flew out at him. The dog could not reach the boy, but the little fellow was so affrighted that he stood motionless for two or three minutes, and at last fell to the earth. He was picked up by some kind passer-by and taken home, and from that moment was stricken by the fatal disease called diabetes, of which in time he died. In this instance there was the direct physico-mental shock followed by physical change, in line. There was the metaphysical vibration of fear transmitted by sight and sound into the body; there was the nervous storm engendered in the body; there was the resultant in a modification of chemical action, by which, in continuous new conditions, a part of the food taken into the body was changed into glucose or grape-sugar; and, on the formation of this sugar in excess, there followed a new series of other organic changes, ending in destruction of that unity of functions which makes up what we call life. I need scarcely say that the illustration above supplied is one in which a mental impression, made through the mind upon the body, was exceptionally severe in its physical effects. But such severe effects have to be seen before the great and primary truths they teach can be recognised.

I was myself many years in practice as a physician before I fully recognised these physical changes wrought through the windows of the mind. It is true I had read of those who were almost bechilled to jelly by the act of fear, but then I looked upon such sayings as mere flights of poetic genius, and in medical literature proper I discovered no clue for guidance in this beat of observation. At last such facts as the one I have stated arrested my attention, and since it has been so arrested I have been daily studying the subject with increasing interest. I could fill this essay and many essays with details of observed phenomena of physical disease from mental action under mental shock.

Indeed, in so many forms do the mental impressions

tell on the bodily organisation, that mental health in education becomes a new branch of science which all persons should begin to learn. By the assistance of this learning our successors will formulate a new world of thought, and will in no small degree fashion, physically, a new world of women and men, having the garb of their souls structurally finer, stronger, and more tenacious of life, from whom shall come a new evolution of species, and a new living earth.

On this inviting theme I must not longer dwell. It is my desire now to treat on those bad mental influences in education which undo the mental and physical health, and on the modes by which these injurious influences may be removed.

Constitutions of Mind.

Suppose we had before us in our schools a body of children all of whom were typical specimens of health. It would then be a momentous fact to know that we could, by our methods of feeding the children with knowledge, make them all specimens of good or bad health. But the truth is that, when we have before us a class of children, we have probably not one before us who is a typical specimen of perfect health. It is a solemn thing to say, and yet it is as truthful as it is solemn, that I have never in my whole professional life seen a perfectly healthy child, and I doubt if one exists in the land. The birthday of health is not yet in the almanack. As a rule, in the majority of children of every class, there is some prepared mode of departure from health inborn in its members. In many of its members the bad health is not merely inborn, but is in actual existence, easily detectable under scientific research. How important, then, that in the modes of training the mind such modes only should be selected as would lead to the better development of both body and mind! How vastly important that all modes shall be avoided which shall lead to a lower development of the mind, and of the body through the mind! If, indeed, it could be that the mind could be elevated while the body was degraded, I, for my part, should doubt the wisdom of education. And if it be really impossible, as I should maintain it is, to elevate either mind or body alone, and absolutely impossible to make one great and the other little, how wide a problem lies before us in respect to education in this age!

What, then, are the modes to be followed in education by which the mental training may be made conducive to both mental and physical development and regeneration? May we think of such modes? I am sure we may, and practise them also. At the same time, the thought as well as the practice requires to be considered from new points of view of an educational kind.

Let me proceed to indicate what seem to be some of the basic changes that must be made in education in order to found a system of mental and physical health on education. I cannot pretend to do more than touch on a few of these changes, the more prominent to my own mind, but far from a complete list.

In the first place, there is, I venture to think, too much friction of mind in education, and, as a consequence, much injury, mental and physical, from cross nervous vibration, owing to the plan which now prevails of treating every boy and girl as if every boy and girl had the same nervous construction and mental aptitude.

As it seems to me, there are as distinctly two grand divisions of mental aptitudes as there are two grand divisions of sex, and any attempt to convert one into the other is a certain failure. The two divisions I refer to are the analytical and the synthetical, or, in other words, the examining and the constructive types of mind.

In our common conversation on living men with whom we are conversant in life we are constantly observing upon them in respect to these two qualities of mind. We say of one man that he has no idea or plan of looking into details; he cannot calculate accurately; he cannot be intrusted with any minute labour of details; but he can construct anything. Give him the tools and materials for work, and he will build a house; but if he had to collect and assort the tools and materials, he would never construct at all. We say of another man that he is admirable at details, and can be intrusted with any work requiring minute definition, but he has no idea of putting anything together so as to produce a new result or effect.

Moreover, we assign to these different men distinctive services in the world. We understand them perfectly, and by an unwritten and, I may almost say, by a spontaneous estimate we reckon them up and give them their precise place in the affairs of life with which they are connected. It is as if by design of nature these classes of men, and it may be of women also, exist as pure types of intellectual form, have always existed, and are always being repeated. In other words, it is as if they are definite families, and that out of them, as out of a dual

H

nature, that human organisation of thought, which we call history, is educed.

The elements of the analytical and synthetical minds appear on a large scale in the pursuits which men follow. The mathematician is analytical, and he, in whatever science his powers are called forth, is always working on the analytical line. He may be an astronomer, a chemist, a navigator, an engineer, an architect, a physician, a painter; but whatever he is, all his work is by analysis. We often wonder at his labour, at his accuracy, at his fidelity. We may say of him that he approaches nature herself in the magnitude and perfection of his results, but we never say of him that he is inventive or constructive. From him much that is quite new comes forth, but it is always something that he has hauled out of the dark recesses: he lays his treasures at our feet, and we are content to admire and wonder. We may be entranced with our view of the produce of this man, but he very rarely kindles our enthusiasm for him as a man, and very often we find that no credit has been given to him as himself deserving of it. We praise only his industry. The poet is, on the other hand, synthetical. This does not always follow, but it usually does, and I think we may fairly say that every man of a purely constructive mind is a poet, albeit we may not be able to say that every poet is constructive. But in whatever particular phase of life and action he exists he shows his synthesis distinctively. His tendency is naturally to drift into such labours as are inventive and constructive. Frequently he avails himself of the labours of the analyst whom he unconsciously follows, believing meantime in himself alone. He makes for us romance in literature;

mechanical instruments in handicraft; pictures in art; tunes and melodies in music; plays and epics and songs in poetry; strategies in war; laws in parliament; speculations in commerce; methods in science.

The two orders of men are often as distinct in feeling as they are in work. They do not love each other, and they admire each other little. Jealousy does not separate them, but innate repulsion. The analytical looks on the synthetical scholar as wild, untrustworthy, presuming, hasty, dangerous. The synthetical looks on the analytical with pity, or it may be contempt, as on one narrow, conceited, and so cautious as to be helpless; a bird that has never been fledged, or, being fledged, has not dared to stretch out his wings to fly.

It has in rarest instances happened that the two natures have been combined in one and the same person. It is, I think, probable that this combination has been the reason for the appearance of the six or seven greatest of mankind. As a general fact, however, the combination has not been fortunate. It has most frequently produced startling mediocrities, whose claims to greatness have been sources of disputation rather than instances of acknowledged excellence.

These orders of mind, distinctive of the distinct, are in their primitive forms so essential to the course of progress, that it is difficult to assign priority of value to either. The analytical mind seems to be most industrious and soundest in practice: the synthetical, the most brilliant, and when on the right track the most astounding, in the effects it produces. The analytical is the first parent of knowledge, the synthetical the second —both necessary.

To apply this reasoning to our present argument, I maintain that, as the child is the father of the man, so in every child there is always to be detected, if it be a child of any parts at all, the type of mind. I will undertake to say that every experienced teacher could divide his school into these two great analytical and synthetical classes. He might have a few who combine both powers, and he would no doubt have a residuum, a true *caput mortuum*, that had no distinctive powers at all; but he would have the two distinctives. He would have scholars who could analyse as easily as they could run or walk, and to whom the mathematical problem and all that may be called analytical is as easy as play, but who have little inventive or constructive power. He would have the scholars whose minds are ever open to impressions from outer natural phenomena, who have quick original ideas, who have, it may be, the true poetic sentiment, but who cannot grasp the analytical and detailed departments of learning at all. The illustrious William Harvey was a scholar of this latter type. It is related of him that late in his life he was discovered studying Oughtred's 'Clavis Mathematica,' and he remarked then that the simplicity of the propositions—their obviousness as it were—had formerly been an obstacle in his way. Harvey was simply a pure type of a most original, and I may go so far as to say mechanical, mind, which, abashed in youth before mathematical problems, in later life, when the reasoning faculty—the wise faculty—was brought to bear upon the difficulties, looked on the understanding of them as difficulties merely from their self-obviousness and simplicity.

The moral I draw from these outlines of natural fact is that in teaching it is injury of mind, and thereby injury of body, to try to force analytical minds into synthetical grooves, or to try to force synthetical minds into analytical. I have an instance under my own observation at this time in which a worthy, a most earnest, and I may add most practical, mathematical master is trying to teach a boy, whose mind is all for construction, the details of the science of details. He had better try to get a third chemical element out of water by chemical process, for that task, hard as it might be, could possibly be a success. But this boy, bright of brightness when the lines on which he can tread are before him, is hopeless here. The master may be angry or perplexed, the parents disappointed;—the thing cannot be done. If fifty masters could be employed in the effort, or if the ability of fifty masters could be forced into one master, the thing could not be done. By a mere act of temporary cram, the thing might be carried out in what we may call a treacherous manner; but it could not be carried out by an honest and reliable education of that youthful mind. Meanwhile, the injury that is being inflicted on the youthful organism is incalculable. Time that could be usefully expended is ruthlessly cast away. Then, the mind itself is rendered irritable and obtuse with each lesson, and the hope deferred makes the heart sick in the truest sense of the term. The failure of each lesson tells on the heart, making that organ irritable and uncertain —making its owner, in fact, 'sick at heart.' This tells in turn on the stomach, causing persistent dyspepsia, and soon there follow the trains of sensations of disappointment, fears of failure in other things, anger at sight of the success of other minds, and all those troubles which lead to the perversion of feeling which so easily becomes the promoter of universal doubt and the opener of despair.

Teachers of youthful scholars will recognise so readily and fully the facts I name, that they will perhaps wonder that I should relate them. Let them pardon me for the sake of the object I have in view. They know, and I know, that these natural differences exist, but the fathers and mothers of children of such differing capacities do not know. The parents look upon all children as alike, and expect all to be turned out of the same brand. If the children are not turned out of the same brand the fault, of course, is the master's, and the master or mistress is thought to be very conceited or overbearing if he or she presumes to state the truth. Perhaps, therefore, it is best for me, who am not a schoolmaster, to speak the truth in all its nakedness. I am only one of the public, and can bear, without harm, any amount of chastisement for my temerity.

As a practical outcome of this part of my argument I should suggest to the public that the members of the scholastic profession should be duly encouraged to try and discriminate, in the case of all their scholars, what is the natural bent of the mind of each scholar; and that, having found this out satisfactorily, they should be further encouraged to train the scholar according to his bent of mind, in order to make him what he really can be as distinct from what he never can be made by any forced attempt at producing the impossible.

Constitutions of Body.

A second point in relation to mental health in education to which I would wish to draw attention relates to the constitution of the body, the stamina of the body—to use a good and expressive term—for work of mind. Just as children of quite different mental stamina are set to the same labours, and are expected to do the same kinds of labour with equal success, so in like manner children of different bodily stamina are expected to do the same labours, and to produce out of them the same results. No error can be more fatal. The class is under the eye of the teacher, in line before him. In one sweep of vision, if the class is a large one, he takes in all the diatheses, all the deep constitutional tints and taints of disease. If he swept his fingers over the keys of a pianoforte he could not detect a more definite series of regular changes.

There is the child with blue eye, light flaxen hair, fragile form, pale cheek, finely chiselled ear, delicate hand, quick apprehension, and nervous, almost scared, nature. That child can be taught almost anything and everything. It may be a very ambitious child, but it is easily put down, and it is always, on the least emotion, vibrating or palpitating. It is the type of the true tuberculous child. You will find of a certainty that some members of its family have died of tuberculous disease in one or other of its forms, most likely of pulmonary consumption. This child may be precocious to an extreme degree, may lap up learning like water, and become morbid in the acquirement of knowledge, but it is always vibrating and constitutionally feeble.

There is another, of the same general construction but of much coarser mould, an obviously defective child with nothing to fascinate: a head probably a little misshapen, the crown somewhat raised and pointed; the face pale; the eye blue or bluish-grey; the ear not well shaped; the hair stiff, so that it has to be cut short to look passable; the hands large and clumsy; the mind rather stolid, and not over-appreciative, but fairly steady at work; the manner subdued and obedient; the nature trusting, but somewhat selfish, and often fretful. This is the type of the strumous child. This child never can work with zest: it has no precocity: when it labours hard, it soon becomes as it were benumbed, and the firmest teacher bids it go out and run, or lets it sit down and sleep.

There is another type in the school equally distinctive. The head is large; the face large and probably ruddy; the lips large; the eye grey or light blue; the hair reddish-brown; the ear large, with a big lower lobe; the hands big; the body inclined to be plump, and the joints large and clumsy. The minds of this type are slow, but at the same time receptive: they are good-natured and heavy, but they bear disappointment badly, and punishment of all kinds *very* badly. Neither much work nor much play is in them. These are types of the rheumatical diathesis. You would find in them, as family physical taints, rheumatism, neuralgia, gout, as direct conditions of natural descent; and epilepsy, chorea Sancti Viti, heart-disease, and dropsy as the secondary or indirect manifestations of the primitive taint which they have inherited.

There is a fourth class, most distinct from all of the foregoing: a type of child in which the body is small;

the head, by comparison, large; the eyes very dark; the complexion swarthy; the hair dark; the lips large; the nose large; the ear large, and the lower lobe pendulous; the body either very small and fragile, or of a size above the usual; the mind appreciative, absorbing, reticent, and self-retained, with a keen sense of its own individual interests, but with small sympathies, and with brooding imagination. This child is a type of the true bilious temperament. It has always in it some blood born of a tropical clime: it has great capacity for work of a mental order, and often for varied work of that kind. It is a type of child fairly healthy during childhood, but suffering often from dyspepsia, facial neuralgia, small eruptive swellings, and frequent depression of spirits, amounting sometimes to actual sadness. It has a very limited capacity for all muscular efforts involving the qualities of endurance and courage, but it is devoted usually to music, and is gifted with musical and artistic ability.

Lastly, amongst the really prominent types, there is the scholar of low mental capacity altogether, and by physical condition incapable of illustrating the active working mind. The children of this type are usually either of small or of very gross build of body. They are unduly pale and fragile; they have irregular or notched teeth; compressed features; very scanty and dry hair; often some bodily deformity, such as strabismus; diminutive heads; and a feeble, sluggish circulation. These constitute, mainly, the class of children whom I have described in my work 'Diseases of Modern Life' as children in whom idleness is a veritable disease. You may do what you will with them, you cannot make them work;

you may pet them, encourage them, punish them, they are the same. They grow up listless and helpless, and, as a rule, die of some organic disease of a nervous character before they have reached the full meridian of life.

I have drawn out sharply five classes of types. In these there are various shades and qualities. In the first class there is now and then a specimen of great mental strength, and often of great physical beauty. In the second, there is often extreme vigour of mind, brightness, and tenacity. In the third class there are, as a rule, many specimens in which both mind and body are active and powerful. In the fourth the mental power is frequently excellent and strangely analytical in its character. Of the fifth I need say no more than has been said.

In large schools with the scholars of which I have come in contact it has occurred to me to observe all the distinctive types and shades of type here named, and a few times in science-teaching I have been able to compare and test in a fair way the mental by the side of the physical characteristic. Those who are teachers know these classes as well as I do, I dare say a great deal better, though they might not like to define them so minutely. I define them because I want to enforce this grand truth, that it is utterly hopeless for parents to expect the teachers of their children to produce great results while the system is enforced of teaching all these children on one uniform system, and while the teacher is debarred the privilege of forming a judgment of capacity in respect to the individual scholar. There can be no mental health in education while pupils of the last class I have named are put in order with those of the first and third. There can be no mental health in education while the brightest and quickest of the first class, the precocious of that class, are allowed to indulge their precocity for learning, and are trained into an ambition which almost of a certainty will, in a very few years, imperil both their mental and their physical organisation.

The practical lesson I would enforce is that the teacher and the parent of the child taught should have between them a better understanding in relation to mental and physical capacities. The quick precocious child of the first class may, under pressure, be taught anything, but the exertion of pressure is at the risk of future disease of the most fatal kind. The child of bilious temperament may be taught with difficulty, but the effort to teach it may be the most useful in rousing its physical powers into new and active life. The first can be killed through the brain, the second can be saved through it. While, in respect to the last-named class, the class of child in whom the brain-cement is so consolidated that there is no free cellular activity, every attempt to overcome inertia may be the very means of increasing and intensifying inertia.

School Work and Overwork.

From the reflections which arise after the study of these different classes of children, I am next led, in thinking over the matter of mental health in education, to touch on the subject of limitation of work in youth. The more I see of school labour, the more certain I become that the strain commonly put upon the youthful mind is altogether opposed to health. It is a matter now of nearly daily task for me to have to suggest relaxation or removal of the young from school or student labour on account of health. In these days no organs of the body are forced so much as the brain and the senses which minister to it.

There are two reasons for this cause of evil action.

The first reason is the utterly absurd general opinion that the period of education is to be limited by the periods of life, and that with the attainment of the majority the day of learning has ceased. If we could get over this transparent yet all but universal fallacy, we should do more to regenerate the world than by any other effort of an educational character. We could then make life a continual feast of learning. We could fill the vacancies between business and rest, vacancies which are now filled often by the most poisonous and injurious pursuits called pleasures—pleasures which satiate by their repetition and ruin by their inanity; we could fill these vacancies with delights of new worlds of knowledge which, ever changing, were ever bringing new spirit and wholesome repose. We should do far more than this—grand as the prospect of cultivating an unwearied life may be—we should take off the strain from the young brain, when all the natural powers are required, not for the using up of the brain in the service of learning, but for the service of the brain itself, for its own growth and development and preservation.

My view is that the duties of the teacher and of the learner in relation to learning should never cease, but that the aim should be to discover in what periods of life such and such processes of learning are best cultivated, and to make life divisible into periods devoted to the attainment of certain phases and forms of knowledge.

I take the case of one I know best. He, when a boy, had great powers of memory for words and discourses and poetry, but had then little power of memory for dates and details. When he was thirty that power of memory by committing to heart began to fail, but the power of memory for details improved in a surprising degree, so that he could without an effort learn new sciences which before were to him closed books. Later on in life he found, in like manner of change, a facility for artistic learning and for the study of forms of which earlier in life he had no notion.

What is true in this one case is, I believe, true of men generally. The man I refer to has, in later life, simply found it easy to acquire that which was not by force forced upon him, and thereby forced out of him in early life, so that in many ways he would actually like to pick up his satchel and go to school again. We want this finding extended generally. If we could take off the pressure of early mental training, so as to improve the mental health by education, we should in turn improve the methods of education. We should do this in various ways. We should limit time so that boys under twelve would not be pressed with more than four hours of work, and girls with not more than three hours, daily. After this we should gradually apportion more and more of time for work, until the maximum of six hours for either sex was obtained.

In other ways we should conserve. We should not strive to teach by short cuts and clever devices until such short cuts and clever devices become more complicate and laborious than the subject itself which is taught by them. I give one example, and that only, of what I mean. There is a book recently published, called a Latin Grammar, in which the Latin language is tried to be taught—for I presume teaching is the aim of its composer—by rules which are, to my mind, much harder to learn than the language. To make these rules facile they are illustrated by doggerel verses so atrociously bad that they make the flesh creep to listen to them. They would have knocked all the verse out of Shakespeare himself had he been tortured with them. The object, I am told, is 'short cut.' To enable many facts to be taught in a short time, it is requisite to artificialise the mind with foreign matters, in order to make it take in more: therefore so much brick rubbish is used on which to lay an unsound foundation for an edifice that is not intended to stand beyond the majority of its owner, but which is fully expected then to fall to the ground or to remain a useless ruin. So the minds of grown-up men are filled with ruined edifices of learning, shapeless, empty, and valueless.

To the errors which are thus cultivated by the crush of education in early life, and which breed a dislike for education in after-life, there is added, in our modern systems, another error, that of making learning, which should be quiet as a murmuring stream, competitively furious. I confess I stand daily appalled at the injury to mental and physical life which I see being perpetrated by competition in the name of learning. Thirty years ago matters were getting bad, now they are getting hopeless. At that time one sex, at all events, was safe from the insanity. Women were saved from competitive mental strain so that the progenies that were to come and replenish the earth were born with promise of safety on the maternal side at least, from mental degeneration. Now, however, women are racing with men, in strife to find out who shall become mentally enfeebled and crippled first. The picture looks terrible indeed.

The picture is terrible, and for the future would be positively calamitous, but for one gleam of hope which, as I will show by-and-by, is cast over it. At this time we look fairly and honestly round to find a great many men still playing an active part in the affairs of this world, writing useful and amusing books, conducting great organs of public opinion, making discoveries in science of the most extraordinary kind, composing songs, and, in a word, keeping alight the intellectual fire. Who are these men? We read their lives, and find that they are, I had almost said without an exception, men who in their early career have been under no competitive pressure: free men whose brains at the period of maturity were not filled with ruined edifices or whitened sepulchres holding dead men's bones. This, we may say, is satisfactory so far. It is. But then comes the solemn question :—Who are to follow these? We look at the past history of men, and see that heretofore the men have always come. We look at the present, and are obliged to say: yes, but in the future where shall they come from? The dearth has commenced in earnest, even at this time. How shall it be removed?

In the upper and middle classes the dearth cannot but remain while the current method of encouraging mental death by competitive strife is the fashionable proceeding. War-cries in learning, as in every other effort, have but one end—desolation, desolation! I am going to say a bold saying—bold because it is based on natural fact. I can find numbers of men who, having been born with good natural parts, have been turned into practical imbeciles by severe competitive strain; but I challenge the production of even one man of pre-eminent and advanced power who has been brought out in complete and sustained and acknowledged mastery of intellect by the competitive plan. 'Glamis has murdered sleep'—competition has murdered mind. There is one university which more than all others is the offender, and the leader in this movement. It is not a teacher; it is a destroyer of teaching. I do not call in question its good intentions, but I oppose its pride and declare its blindness; and I want all who are engaged in education to protest against the ruin of the good work which it and all who go with it are inflicting so determinately.

I said I would light up this subject with one gleam of hope for the future. I take that gleam from the Board schools; it is kept in them, and I trust it will be always. If the Board schools will only maintain a moderate system of education; if they will simply be content to lay the foundations for the development of such men as Shakespeare, Priestley, Fergusson, John Hunter, James Watt, Humphry Davy, Michael Faraday, William Cobbett, Turner, Flaxman, Richard Cobden, Charles Dickens, George Stephenson, David Livingstone, and others of such sort, all of whom would almost surely have been mentally destroyed by the competitive ordeal, they will do a work which will be more than national, a work world-wide and lasting as time.

Haply, too, in the success of their undertaking, the Board schools may, by force of results, bring back to reason the erring crew who would cram all learning into the human mind in the first quarter of its existence, and leave it stranded there. It is a sad look-out for the now governing classes, one million in twenty-four millions, if this lesson be not soon learned. For knowledge alone is power, and knowledge with wisdom combined is victory and governance.

In this suggestion for the future, no thought is conveyed of placing the Board schools in opposition to the higher-class schools and the Universities. The higher-class schools and the Universities of these islands have played, in the past, a part second to none elsewhere. They have had their princes of knowledge, their Newtons, their Halleys, their Hamiltons, their Harveys, their hundreds of great scholars, poets, philosophers, all that is mentally noble, as their own. My argument is, that these great ones were theirs when they were content to cultivate industry, to nurse genius, and even to fan into life what might at first seem feeble and unpromising mental effort;—but that the like of these can no longer be theirs, if they continue to care less for true culture than for the apparent, and only apparent, results of culture; and if, instead of sustaining the weak, they strive to become powerful by crushing and killing in their early life the strong as well as the weak by the like impatient pressure.

I had intended to touch on education as it should be modified according to seasons of the year, and on one or two other equally important topics; but my time is up, and I therefore content myself with offering, as the essence of my discourse, the following propositions:—

1. To secure health through education, it is requisite that a more systematic and scientific study of the psy-

chology of the subject should be undertaken, and that class studies should be divided in regard to the mental aptitudes of the scholar.

2. Parents should expect teachers to exercise a fair and discriminating judgment as to the particular capacities of children under their care, and should be influenced by such judgment in the direction of educational work. The teacher should become, in short, like a second parent to the scholar.

3. Much greater care should be taken in observing the influence of special physical peculiarities of body and heredities on educational progress, while the influence of education on such peculiarities and heredities should be carefully learned and determined. By this means two useful purposes would be secured: education would be made to conduce to physical health, and physical health to education.

4. All extreme competitive strains in learning should be discountenanced as efforts calculated to defeat their own object, and to produce mental as well as physical degeneration.

5. In school-work, the Swiss system of teaching should be more closely followed: that is to say, very quick and precocious children should be directed rather than forced and over encouraged, while dull and feeble children should receive the chief attention and care of the teacher.

6. Education should be so carried out as to make the whole of the life of men and women a continued process of learning, varied, at different ages, according to the changing capacities, faculties, and aptitudes for the different subjects included under the head of knowledge useful and universal.

NATIONAL NECESSITIES AS THE BASES OF NATIONAL EDUCATION.[1]

WE have been discussing, for some weeks past, at the London School Board, the question of higher education, and after many debates have not, as yet, arrived at a satisfactory conclusion. In the course of the debates, two contending principles have been brought out. On one side it has been enthusiastically declared that teaching of a higher standard than that which is now common is necessary, both for the teacher and the taught, because the teacher feels the tax of one continual grind on elementary subjects an intolerable burden, and the taught fail to receive what might, in many instances, prove to be the inestimable blessing of a superior education. On the other side, it has been urged, with great earnestness, though, of course, not with enthusiasm, since enthusiasm can only be allied to aspiration, that the business of the Board ought strictly to be confined to the objects of providing a plain and elementary education for the many thousands of pupils it has under its care; that the development of higher-class teaching should

[1] Revised and enlarged from a Lecture delivered before the Society of Arts on April 28, 1882.

rest with those who have the means of paying for it; that if the basis be laid for sound elementary instruction, all who desire to obtain a better-class education will, themselves, find the means; that, practically, the present system gives the scholar the key by which he may open the door leading from the regions of darkness to the region of light, so that, being in the light, he can go whithersoever he will; and that, as a consequence, every attempt to add more teaching in the elementary school is a departure from economy, and a misappropriation of the funds which the members of School Boards hold in trust for the public.

I have taken no part in this discussion, except to listen attentively to it, and try to extract from it that which seemed to be useful; and the lesson I have learned is, that in certain ways both parties concerned are in the right and in the wrong. I entirely sympathise with those who say that the present labour of the schoolmaster and schoolmistress must needs be a burden that becomes a daily cross; must be disheartening to a degree that those of us who are engaged in varied pursuits can scarcely recognise; and, though very grand in its results, must be as disappointing to those who are engaged in it, as the mere laying of the foundation of a grand cathedral pile must have been to those who failed to live to see the structure rise beyond the ground, and become what it ultimately would, the admiration of hundreds of succeeding generations.

I am not without sympathy, at the same time, for those who reason on the economical side. I agree with them that when a child of fourteen years can read well, write well, and calculate well, it has done as much as it ought to be allowed to do, in that way, up to that age. If it be forced to do more in form of brain work, it is forced to do what is physically wrong for its body's sake, so much power of work required for its nutrition having been extracted simply for the development of mental aptitude and accretion. I am quite sure, indeed, that in a future and a wiser day, when the physics of life are better understood, men and women generally who have determined to live the whole term of life instead of one third of that term will not care for their children to be troubled with book-lore at all, previous to that important physical stage of life marked out by the first of the seven stages which ends between the fourteenth and the fifteenth year. I sympathise with the economists on yet another ground, namely, that to prime the young with the idea that they are only to learn while they are young is to crystallise them into old men and women from their first, and of a certainty to shorten their lives; because learning is as necessary to perfected life as bread, and because the happiest human existences are the existences of those persons who are always slowly acquiring knowledge in its endless variety of form and character, and who, as they grow older, apply what they acquire the more wisely, effectively, and satisfactorily.

Why, then, it will be asked, if you feel these views, have you not used your right to express them from your place at the School Board? I answer at once, that my views would not have been in order, if expressed, inasmuch as they would not have related to the subject actually under debate. The question under debate has been for more book-learning, and more expense, by the first section; and for no more book-learning, and no more expense, by the second section of speakers. I do not sympathise with the first on the matter of increased book-learning; I do not sympathise with the second in favour of mere economy for economy's sake; and therefore I could not expect to be considered in order on the particular subject under discussion.

There is, however, another subject which no School Board has touched, and which no School Board can touch, until the public mind has become familiar with it. I refer to the question whether the present system of national education is based on the national necessities. In a few years this will be the leading topic of the School Boards everywhere; at present it is in that embryonic condition in which it can only expect to be tended and nurtured by such a society as the Society of Arts, which, as our late distinguished and lamented colleague, Sir Henry Cole, once told us in Council, has for its first duty the duty of becoming the nursing mother of new and useful products of thought, until the nation takes them up, and puts them into business on their own account.

It is for the purpose of bringing forward this new aspect of the educational question that I stand before the Society to-day. I ask myself if the system of education at present going on in our nation is a system which has a proper relation to the necessities of the nation. I look round me to see the nation in chaos of thought and action: in what Mr. Gladstone has correctly defined as social revolution in all parts; a revolution that might, by the merest accident, by one or two days' shortness of food, from failure of foreign supply and panic thereupon, pass, after a few weeks of further chaos, into physical revolution. The thought which occurs to my mind, as it must to the minds of all who think, is, are we educating to prevent catastrophe ? Are we educating the young to become useful, independent, and capable working members of society, ready to work with muscle as well as brain, in orderly and profitable form ? Or, are we educating them to become mere troublers without design ; repiners without hope ; schemers without self-control ; masters of the forces of nature herself, and knowing how to use them for temporary, selfish, insane objects, but not knowing how to apply them for splendid purposes and for the general good ?

Let me at once say, that this apprehension, and, I may add, comprehension, of a great national question is not mine. It was advanced by Andrew Combe, it was enforced by George Combe, and before either of these great men wrote it was formulated and insisted on by Gall and Spurzheim. But the man who has the longest and the most earnestly advocated this view, urged it, pressed it, in season and out of season, for forty years or more, and who has held it up to this hour unflinchingly, is Edwin Chadwick.

The national necessities, as the bases of national education, call, first and foremost, in the early days of youth, for the three simple elementary educational practices of learning to read, to write, and to calculate. But these necessities are comparatively valueless unless they be combined with further necessities of a physical kind, namely, sound and systematic muscular exercise ; freedom of breathing and circulation of the blood ; practical training, so that the body can be structurally builded up and sustained in health ; preparation for all duties requiring precision, decision, presence of mind, and endurance; and readiness to acquire any craft or handicraft that may lead to a useful living. In a word, an education that shall bring the mental and physical qualities of every person into faithful harmony and goodwill.

It will be clear from these expressions in what way I agree and yet differ, remain with and yet stand apart from, my colleagues at the School Board in their recent discussion. I, like some of them, would break up the monotony of the schoolmaster and schoolmistress, and would give those excellent workers as much variety of teaching as any of them could desire. But that variety should be physical, not mental, play rather than work; training of the muscles, and, I may say, of the skeleton too, of the lungs, of the heart, of the digestive organs, and of brain and nerve for action; not training of brain alone, all the day long, but of all the organic parts equally. I, like others of my colleagues, would encourage economy, not by keeping things as they are, but by saving some part of the two-fifths of the money now expended on teaching to spell, and by laying it out in teaching how to walk with grace and ease, to sing with correctness, to swim, to learn the use of the arms and fingers and hands, and to become men and women in the strict sense of the word, without danger of retrograding a hair's breadth in the Darwinian line.

I said in my address at the Health Congress at Brighton what was quite true, that I had never in my life seen a child so healthy that it had not in it some actual or latent constitutional disease. Touching the subject now in hand, it is equally true to say that it is all but impossible to find in the Board schools of our large towns any semblance, critically viewed, of health. Constitutional taints, which under favourable circumstances may often be concealed, and which may or may not be apparent, are there. Various conditions of disease are there independently of the tendency from heredity; there of themselves, in some irregularity of body or limb, in some imperfection of sense, in some deficiency of quality of blood, in some feebleness of respiration, in some nervous irregularity of function, in some shade of mental aberration.

The field of disease which is presented in some of the schools situated in crowded localities is indeed a sight at once for anxiety and pity. To the eye of a physician who, like myself, has spent many years in hospital practice, it tells a story which is absolutely painful, if he permits the result to be calculated out in his mind at leisure hours; if, that is to say, he compares what he has witnessed in his survey with what he has learned from long observation of the meaning of the phenomena in the history of life. It is not necessary for him to strip the children, percuss and sound the chest, examine the spine, or practise any of those refined arts of diagnosis with which he is familiar. He reads from the indications of temperament, of expression of countenance, of colour of skin, of position of limb, of build of body, of gait, of voice, sufficient outward manifestation to discern what is the true physical state, what is the stamp and extent of disease, what is the vital value of the lives, generally, that are before him. Tell the physician those lives are to be valued for some monetary purpose as they stand and as they are to go on according to the present system, and he will give, in brief time, an estimate of value which the keenest man of business might readily accept and act upon.

Foremost amongst the evils which are thus presented are those common conditions of disease known as anæmia and cachexia. Strictly these are not diseases, like diabetes, bronchitis, or defined affections running a regular course, but they are states of disease which by their very presence indicate a faulty nutrition at the period of life when good nutrition is most required, and which cannot long go on without ensuring the construction of an impaired bodily organisation. The blood is not being duly oxygenated, and food, therefore, though it be even fair in quality or quantity, is not properly applied. The nervous system is imperfectly built up; the skeleton is imperfectly built up; the muscular system is imperfectly built up and sustained. How can the improvement which is called scholarship be turned to fitting account in such recipients of it? I watched recently the afternoon working of a large class of scholars, and counted one third of them under the most decisive influence of these conditions of disease. Of the affected there would not be, in the ordinary averaging of life, twenty years of existence under the course that was being followed. The one saving clause in their case was development by physical training, and that was withheld. The one destroying clause in their case was over-mental work without the physical training, and that was assiduously and regularly supplied.

With or without the anæmia and cachexia, there is the constitutional disease, struma or scrofula, presented in these classes. The instances of this kind in varying degrees of intensity are most numerous. This condition again is a mal or bad nutrition. It, as much as cachexia and anæmia, with which it is so often allied, is fostered by the prevailing system of mental pressure.

With these conditions before the eye, there is to be seen, also, here and there in the classes of both sexes, but of the girls especially, the specimen of the phthisical or consumptive subject. In a class of fifty I pick out three thus doomed, if their circumstances be not changed; six per cent., certainly a moderate proportion. The disease has not positively developed, but the probability of its development is all but certain unless it be checked by the one only remedial or preventive method, freedom from nervous exhaustion combined with physical exercise in open breathing space. Such preventives are not supplied, but undue nervous exhaustion and confinement are both supplied, and so the fatal disease is systematically fanned from latency into activity.

Spinal deformity and irregular construction of the skeleton is another condition of disease, or actual disease, readily detectable in these classes.

Miss Löfing, speaking of her experiences as to the girls which have come under her notice, reports what is but too true, that they are, as a rule, very flat-chested, that there is evidently much spinal distortion, and that lateral curvature of the spine is common amongst them. This—which, I may state in parenthesis, is equally true in respect to boys—is accounted for by Miss Löfing in terms which show that the present school system does more than simply permit the mischief that is progressing, it actually fosters it and promotes it. Asked to what the effects are chiefly ascribable, she replies :—

'A part is ascribable to home neglects; but the

greater part is due to excessive constraints under the common school conditions: too long sitting on badly constructed seats (a fault which is in course of correction); and even on good seats, when they are kept in bad positions in long writing exercises. The common bad position is, indeed, prescribed by the Government School Inspectors. I have found that, to obtain the school grants, the children are so constrained as to exclude the exercises that are needed for their bodily development. For example, I visited one school where there was a class of children (girls), who were stated to be of an average of about eleven years of age; they were generally of a pale sallow complexion, though they were from a good neighbourhood. I asked the schoolmistress whether these girls had any exercises; she said, 'No, they had none—they had time only for the preparations for the examinations.' Did she, I asked, give them any play? 'No, they had no time for play; they had only time for the preparations.' She said that if she could make two weeks out of one she might then get time for exercises. Another observation she made, upon my suggestion that they really ought to have some exercises, was that exercises would be difficult, as the girls had stays on. These little girls of eight years of age with stays on! The fault is want of knowledge, in the School Department, of the bodily constraint imposed by the preparations for these examinations, and that pressure is really now being extended to the infant schools.

It is right to put these matters of fact prominently forward, and it is quite easy to do so, without suggesting from them that the evils are worse than they were in some past times. In truth, things are better than they were. It is infinitely better for these children to come out of their wretched homes into the schoolroom to be educated than to remain in their homes all day; and I am not, for a moment, contending against the school system on the argument that the children should not come to school. My argument is, that the children, being at school, should have all the benefits derivable from the supervision under which they come, and nothing but the benefits.

Mr. Charles Roberts, in a letter which he has addressed to Mr. Chadwick, puts this question forward in words I should entirely indorse. 'I may observe,' he writes, 'as the result of my study of this subject, that there is a little exaggeration in the prevailing opinions on the rapid degeneracy of the physique of our labouring classes in towns, and that it is not nearly so great as it is often represented to be. The great improvements in sanitation, which you must look on with special satisfaction from the part you have taken in bringing them about, and the recent improvements in the medical treatment of children's diseases, have more than counterbalanced the injurious effects of town life. I think, however, the evil of the physical degeneracy of our labouring population is a very urgent question for the present, and the immediate future. I think children are being very cruelly used by the Legislature. It took nearly fifty years, and half as many Acts of Parliament, to emancipate them from the injurious effects of excessive physical labour; but the result has been merely to transfer them from one taskmaster to another—from the manufacturer and their own parents to the schoolmaster; and to subject them to mental strain and physical inactivity, more injurious to their future well-being than their former condition. I look on your proposal to introduce the half-time system into schools as a sort of mental " Factory Act" of the utmost importance and urgency, and the establishment in elementary schools of systematic physical education as absolutely necessary to prevent great physical degeneracy in future.'

These are some of the serious evils which attend the present school system. I have dealt only with the most important, and I am glad to say that many minor evils which are talked more about, such as short sight, strabismus or squint, club-foot, and other local deformities, are really not so prevalent as is generally believed. There is sufficient of gravity without taking them into our consideration at this moment.

In a minor degree the health, not only of the scholar but of the teacher also, is imperilled. We know well now that, in the adult stages of life, the constant sedentary occupation, capped by the monotony which springs from the perpetual following out of one pursuit, cannot fail to be attended with premature degenerative changes of tissue, leading to premature nervous enfeeblement and old age. On this point the more thoughtful of the school teachers are themselves agreed; and the school-teachers of Scotland, who can hardly be surpassed in practical shrewdness and knowledge, have long declared that the present methods of teaching introduced by the Code are injurious, mentally, bodily, and morally.

We have quite recently had before us another exemplification of these facts in an effort which has been made in the School Board of London to introduce a systematic and, I had almost said, professorial plan of education for pupil-teachers. In this trial, which is as yet limited in its operation, teachers are nominated to instruct classes of pupils in special subjects, the professed teacher keeping to his own particular department of instruction. The pupils, including representatives of both sexes, meet at certain hours regularly each week in class for this systematic work, as they might meet in a college or university, and at first sight one would assume that no better plan could be followed. I was, I confess, much taken with it when it was started, and at a largely attended meeting of school-masters and -mistresses I spoke in favour of it. I am in favour of it still, but, at the same time, I am sorry to say that almost as soon as it has commenced to pass into operation it has commenced to show the cloven foot of the evil of mental overstrain and physical exhaustion. Already the learners are entering into competition against each other and against that much more formidable enemy, the *edax rerum*, time. Already the health of the learner is being interrupted. Already, if I am correctly informed, it is felt necessary to encroach on the rest of Sunday in order that the lessons of preparation may be duly learned. Already it is complained that certain of the pupils who have entered at the beginning of a year are behind those who are a year older and more advanced, and have, therefore, to put on extra strain in order to catch up those who have got a long start. And, already, some of us who have control over this work are beginning to ask if we have not once again reckoned without our host, and whether we must not revise a treacherous but well-meant policy.

To return to the children. The present system is, in

their case, not only a violation of physical but of psychological law. The powers of receptivity of the minds of children of different ages have been tested, experimentally, with as much care as physicists take when they are treating in their experiments on the relationships of ordinary matter to force. You take a Leyden jar of a certain size, and you say you can charge it in a certain time from so many turns of the electric machine; beyond that you can do no more; you have reached the powers of receptivity of that jar. You can break it if you will, but you cannot naturally surcharge it. The powers of receptivity of the minds of children of different ages have been tested quite as carefully, and it has been found that the capacity for attention and retention is as measurable as in the physical experiment. Certain brains can take in so much, and no more, according to age. The capacity grows with cultivation and skilful teaching, no doubt, but it must be permitted to grow. In the very young a lesson of a minute may be all-sufficient; later, of three minutes, five, ten, fifteen, and so on, to one hour, two, or three. But to this there is limit, and it is probable that, with the best scholar of primary school age, the powers of receptivity rarely extend beyond a period of two hours and a half of continuous teaching. Teachers of various districts, and of different countries, have testified in respect to this point, and while they have explained, from direct observation, that the receptivity varies in different children according to difference of temperament and physical health and build, as might very well be expected, the receptivity at one time, in all children, ceases at the end of three hours.

This view is fully supported by Mr. Chadwick, who,

in his essay on the physiological and psychological limits of mental labour, made to the British Association for the Advancement of Science in 1860, reported as follows: 'In one large establishment, containing about six hundred children, half girls and half boys, the means of industrial occupation were gained for the girls before any were obtained for the boys. The girls were, therefore, put upon half-time tuition, that is to say, their time of book instruction was reduced from thirty-six hours to eighteen hours per week, given on the three alternate days of their industrial occupation, the boys remaining at full school time of thirty-six hours per week—the teaching being the same, on the same system, and by the same teachers, the same school attendance in weeks and years, in both cases. On the periodical examination of the school, surprise was expressed by the inspectors at finding how much more alert, mentally, the girls were than the boys, and in advance in book attainments. Subsequently industrial occupation was found for the boys, when their time of book instruction was reduced from thirty-six hours a week to eighteen; and after a while the boys were proved, upon examination, to have obtained their previous relative position, which was in advance of the girls.'

I recall, too, how in the same report the exposition of this practical truth is sustained by my illustrious countryman and friend, Professor Sir Richard Owen. So clear, so terse, so true, is this exposition, I make no apology for presenting it again to the public, after its long rest of over twenty years.

'All the nutritive functions and actions of growth proceed,' says this master, 'more vigorously and rapidly in childhood and youth than in mature life, not merely as regards the solids and ordinary fluids, but also in the production of those imponderable and interchangeable forces which have sometimes been personified as nervous fluid, and muscular force. Using the latter term to amplify my meaning, the excess of nervous force is in the child most naturally and healthily reduced by its conversion into muscular force ; and at very short intervals during the active or waking period of life the child instinctively uses its muscles and relieves the brain and nerves of their accumulated force, which passes, by the intermediate contraction of the muscular fibre, into ordinary force or motion, exemplified by the child's own amusements and by those of some object or other which has attracted its attention. The tissues of the growing organs, brain, and muscles, are at this period of life too soft to bear a long continuance of their proper actions ; their fibres have not attained their mature tone and firmness. This is more especially the case with the brain fibre. The direct action of the brain, as in the mental application of learning, soon tires ; if it be too long continued, the tissues are unhealthily affected ; the due progress or growth, which should have resulted in a fibre fit for good and continuous labour at maturity, is interfered with ; the child, as an intellectual instrument, is to that extent spoiled by an error in the process by which that instrument was sought to be improved.

'The same effect on the muscular system is exemplified in the racers that are now trained to run at two and a half or three and a half years old for the grand prizes at Doncaster or Epsom. The winner of the Derby never becomes an Eclipse or Flying Childers, because the muscular system has been overwrought two or three years before it could have arrived at full development, which development is stopped by the premature over-exertion.

'If the brain be not stimulated to work, but be allowed to rest, and if at the same time the muscles be forbidden to act, there then arises, if this restraint be too prolonged, an overcharged state of the nervous system. It is such a state as we see exemplified in the caged quadruped of active habits, when it seeks to relieve it by converting the nervous into the muscular force, to the extent permitted by its prison, either executing a succession of bounds against the prison bars, like the agile leopard, or stalking, like the lion, sullenly to and fro. If the active child be too long prevented from gratifying the instinctive impulse to put in motion its limbs or body, the nervous system becomes overcharged, and the relief may at last be got by violent emotions or acts, called "passion" or "naughtiness," ending in the fit of crying and flood of tears.'

It would be impossible to add a word that would emphasise this graphic and remarkable exposition of natural law. It is a picture as truthful as it is forcible, and it forms a truly fitting conclusion to the argument I have so far made it my endeavour to set forth, as the physiological foundation or institute of educational improvements.

Proposed Reforms.

Let me now turn to the reforms which we, who are urgent to reform the present system, have in view. I have conveyed all along in this discourse this current thought, that the present system requires a radical change, in which the physical shall at least halve the time with mental education. For such change there is, we maintain, a reason and a method. The reason is the first point to be adduced. We reason that the existing system is not a basis for the national necessities. We are of opinion that in the future the education of a mental kind now being supplied will be imperfect and doubtful, nay, may be of dangerous use, unless it be so laid out with physical culture that a perfect, or comparatively perfect, health of body shall go with it and sustain it. We urge that, as we must either educate health or disease, it is better to educate health.

The design we have in view, then, includes several heads, which we say embrace no more than is absolutely necessary as bases of the national necessities. I may arrange these heads in the following order.

Physical Culture of the Body.—We urge that education should be so distinctly physical that the body should be in no respect less improved than the mind at the close of the educational career. We follow, in this regard, the teaching of the Platonic philosophy, in which the master insists that the symmetry of mind and body be cultivated and maintained, without which there cannot be beauty, there cannot be health. We urge that this is the only sure way of keeping up, in our country, a strong and vigorous and independent population, that shall understand how to utilise the home resources of land and industry, and keep the land and industry in the possession of our and their descendants.

The system of education that is now being carried on seems to us to promote in no way whatever this necessary intention. In the 'Standards' we find no efficient instruction of a technical kind that even in the barest hypothetical style teaches the principles of useful arts and appliances. Practical details of industries and of modes of learning industrial occupations are thought to be of less importance to the scholar than a knowledge of geography, construction of language, physiology, and history. It is no wish of ours to ignore studies of the kind above named, but we consider that elementary instruction in details of inventive and industrial pursuits holds a first place, and that in a country like ours in which so much, in which, in truth, almost everything, depends on individual perfection in some useful art, such elementary instruction ought to have the place it deserves at once and for good. We think, moreover, that the instruction should not be purely theoretical. We contend that it should include elementary training in useful work of a practical kind. We do not see why workrooms should not be set up in schools, in which boys should be taught the use of the lathe; the beautiful art of wood carving; the skill of the draughtsman; the method of distinguishing metals, and other simple experiments in chemistry; the arts of swimming and riding; the art of distinguishing colours and signs at a distance; the practice of mensuration, and a number of other good and useful branches of physical learning, which, whether the boy remain at home or whether he emigrate to another country, will always be to him a direct assistance, a means of earning his bread, and an insight and test of his particular ability or aptitude for the vocation by which his subsistence will be most easily obtained.

Extending this principle of practical teaching to the

female sex, we would have the girls well trained, both theoretically and practically, in those occupations which in the course of life fall more distinctly under their exercise, management, or supervision. We know that in the schools at present girls are taught sewing and a few other useful industrial accomplishments. We would extend these instructions. We would have the girls instructed in modelling; in the art of colouring and painting on glass and porcelain; in the various processes of selecting, sorting, preserving, and preparing foods for the table; in the cleaning and ornamentation of drawing-room ornaments, and in all the works pertaining to domestic life. The girls in our schools would, as we believe, make more rapid progress in mere book learning if one half of the time now devoted to books were devoted to that other branch of practical education which forms the greater part, in practice, of their future womanly life. We consider that evidence in proof of this belief has already been offered, and we suggest that a girl trained in the manner now described would, in this country, or in any other country into which she might emigrate, be far better fitted for the duties pertaining to any station she might hold, than if she were simply dismissed from school primed with the standards, and standardless.

Life-learning Tendencies.—We contend, secondly, that the education of the young of all classes, and of the poorest classes chiefly, should be so framed as to lead to the inducement of making the acquisition of knowledge a taste instead of a task, a liking instead of a labour. We contend that in the present system as it is pursued, children who are not by heredity born to mental occupation, or who are not physically constituted to acquire information, are, by sheer force, driven through the hard and fast lines, fenced out by books called standards, at a pace that tries to make them complete their education irrespective of temperament, health, ability, before their thirteenth or fourteenth year, and that the pressure, amounting in every case to a hardship, has merely the effect of causing them to cease to learn when the pressure is taken off. We insist that this present system should be so modified that there shall be no mental pressure at all, but a mixture of mental and physical teaching which shall bring the mind into desire for knowledge after it is freed from the forced necessities of acquiring it by routine under pressure.

Our argument on this point rests on history as well as on natural observation or common sense. We have from history no proof at all of greatness of mental quality developed by the plan of forcing the young mind through close and wearisome grooves of artificial learning. We see in past history that the greatest examples of greatness in the arts and sciences, in literature, and in skilled labour of all kinds have sprung from those who have been least constrained and least trammelled, in their early days, by artificial and forced systems of education. We see, in current history, that under the favoured and false systems now prevailing in our middle and rich classes, the worst consequences follow upon such artificial and forced educational devices. We see men and women in their early age crippled for good; crippled, often physically and very often mentally, by the cramming ordeals which they are made to go through. We see in them, moreover, the dearth and barrenness of the mental field; the dead mediocrity of talent; the weariness of effort towards new and original advancements; and the surfeit of learning which so unhesitatingly leads the pressganged scholar to accelerate his emancipation from the school or university by the determination that now he has got through his trials he will be free from all such work again, and will fill up the rest of his short life with other and less irksome pursuits, though they may even lead him, as his seniors assure him they will, to *ennui* and monotonous old age.

Seeing, then, the practical effect of the forcing system upon those unfortunates on whom it has been tried, we are anxious to save those from it upon whom, fortunately, it has not as yet been so effectively practised. We would like to see the scholars of the State such scholars that, when their days of compulsory education have passed by, they are anxious still to learn, and are desirous to realise a state of life which makes life one continuous happy school-day, each day a day in which some lesson, however small, has been acquired, and has been added to the satisfaction of accomplished desire. Under such a *régime* I venture to think that the Board schools might become evening colleges, in which the older population could meet to pass through the stages of a steadily advancing and higher education; in which the school professor should supersede the tapster and gin-seller; and from which there should constantly be given forth scholars who should make their schools the friendly and honourable rival establishments of the higher schools and of those great seminaries of learning—the ancient Universities themselves. We feel that under this rule the school would be a place to which the scholar would be attached, and that the feeling of attachment, increasing into pride of the institution in which he was educated—a sentiment which has never, up to this time, been engendered—would be dominant proof of the triumphs of a sound national educational existence.

Aptitude for Productive Ability.—A third advancement upon which we lay great stress is, that the educational system shall be of a kind which shall render the body of fitting aptitude for productive ability. We argue that unless discrimination is used by the teacher for detecting the natural, or, to speak more correctly, hereditary capabilities of the scholar, there must be failure, in results, of the most serious kind; failure that will tell upon all the productive industries of the country, so that agriculture, the various industrial arts, the various labours which call for muscular skill, activity, and endurance, will be sacrificed, or largely reduced in effective value. On this topic I have before me a letter written to Mr. Chadwick by Mr. Isaac Ashe, a Fellow of the King and Queen's College of Ireland, from the Central Criminal Asylum, at Dundrum, Dublin, over which establishment the learned writer presides. I do not think I can do better than quote his exact words on this part of our subject.

'Persons,' he observes, 'used to continued brain-work, and who have an inherited capacity for it, ignorantly compare the standard of mental toil attainable by themselves or their children with that attainable by the descendants of generations of manual labourers. But they might as reasonably set their own children, at ten or twelve years of age, to undertake the continuous labour of a navvy's son of the same age. I have no doubt but that the system of half-time schools which you advocate is the true solution of the problem of how to train to intellectual tasks the dull minds of the humbler classes. It is easy to see how the healthy discipline and regulated training in some artisan's trade or handicraft will save from crime those whom mere intellectual taskwork will surely relegate thereto. To work healthily, either with mind or body, is a habit; and brain or muscle acquires the habit of using up the energies of the body in a particular direction. Such a habit, when once developed, becomes hereditary, and constitutes in any individual's descendants a capacity for a particular kind of labour not readily to be broken through or set aside. If the young are forced to break altogether with their hereditarily acquired and normal tendencies and developments while yet they have been unable fully to mould themselves to a different rule, it is easy to see that their development will be abnormal and unhealthy. But the habit of and capacity for manual labour is necessarily much more easily acquired by the children of generations of manual labourers than are the habits of and capacity for brainwork. Hence to develop and direct into healthy channels and courses of operation the muscular energies of such children must always be an easier task, and one more congenial to them, than that of developing a new capacity— to them almost a new function of brain-tissue—that, namely, of studying and thinking.

'The hereditary tendency of the children of criminals to relapse into crime must also not be overlooked. The question is,—How can it be eradicated? This can probably be effected only by directing into healthy channels the nerve and brain habits, which will otherwise act spontaneously in morbid and criminal directions; and the channel into which these can most readily and with most ease be directed is undoubtedly to be found in something which will call into action the same mental and physical aptitudes as those which the hereditary career of crime shall have developed. If, for instance, the child of the clever forger is taught draughtsmanship, the hereditary proclivity to a criminal use of an instinctive faculty will be directed into an analogous yet healthy channel, and probably with the result of at once curing the tendency towards crime, and turning out a skilful artisan. If the children of generations of pickpockets are taught to use their criminally deft fingers and delicate touch in some handicraft requiring a special capacity of finger, such as watchmaking, or filagree work, a healthy function will be found for a nervous proclivity and muscular aptitude, which will otherwise fairly work itself out in the criminal acts to which its very existence forms an almost irresistible temptation. The children of the burglar might be found, similarly, to have an hereditary aptitude for a blacksmith's trade, and so on. And thus criminality will be eradicated by directing the hereditary faculty into a healthy and normal channel of exercise. But to attempt to abrogate it utterly, or eradicate it as a criminal tendency, without such utilisation of it in a healthy direction, will prove futile.'

Naturam expellas furcâ tamen usque recurret.[1]

---

[1] A passage I would freely translate—
Though man may check Nature by matter of force,
She will take her own way as a matter of course.

In insisting on this practice of developing productive ability, so ably stated by Mr. Ashe in the above remarks, we are sustained by the belief that nothing could be lost by the effort in the way of actual education. We are of opinion that the time saved by the adoption of varying conditions of school work would prevent the injuries now incident to the fixed rules under which the educational system is enforced, and in this view we are supported by the opinions of the most practical teachers.

Mr. Imeson, late of the Central District School of London, whose authority, based on extensive practice, stands high as a teacher, objects to the Code that it has rested erroneously on rigid lines of demarcation. He observes as follows:—'The mental endowments of children are so varied, that great elasticity of method is needed in teaching them. This the inspector as a rule ignores, being bound to a cut-and-dried rule of inspection. Whatever is likely to develop intelligence may be commented upon in Bluebooks; but the ordinary tests of the inspector are mechanical, and made to catch the unwary, overleaping the proper course to be pursued. There is a saying, 'As is the master so is the school.' This used to be true, but now it might be said, 'As is the inspector so is the school.' This is mischievous beyond ordinary observation, for the inspector generally owes his position to his attainments, which are presumed to be of more avail in a school than the teaching skill of a *bonâ fide* education.'

Physical Training and Mental Habits.—We maintain that courses of physical training, such as we wish to introduce, would have a distinct formative effect on mental habits. In a draft report for the consideration of the Education Committee which has been presented to the Council of the Society of Arts, and in the framing of which Mr. Chadwick took the leading part, it was stated :

'Nearly the whole of the industrial and of the reformatory institutions are now conducted on the same principles of mixed physical and mental training as the district half-time schools. Although the children, generally of the lowest type, are received after being already much hardened against school influences, the success of the tested outcome of a great majority of them appears to be upwards of eighty per cent. Governors of prisons at the Social Science Congresses have expressed gratification at this progress, and congratulate themselves that of the usual large contingent of this class of prisoners they are now, as a class, almost relieved. It is important to note and consider the distinct formative process and effect of the physical and mental training in these institutions, by which incipient criminal habits are eradicated, as contrasted with the courses of education in the common schools which have no systematised physical training.

'In the long-time schools, during the time the boy is kept waiting under restraint, his mind is absent from his lessons, which are commonly so uninteresting as to be repugnant to his voluntary attention, and his thoughts are away on cricket, or some sort of pleasurable play, so that he generally only returns, upon call to the lesson, as to a task to be got rid of. Under the restraint of separate confinement in a prison, the mind of the young criminal cannot, as is shown by his action on release, have been occupied with compunctious visitings of remorse, as commonly assumed. His thoughts are of his ill-luck under the wide chances of escape of which he has had experience during all the time he has been at large before detection, and of how he may have better luck when he gets out. He is exhorted to be good: but the child of the mendicant or of the delinquent does not see his way to doing other than he has done before; and why should he so long as he feels his inaptitude of hand and arm for industrial work? Be this as it may, under the common conditions of restraint in the district schools, or in the reformatory schools—all of which, comprising some thirty thousand children, are now of necessity conducted on the half-time principle of varied, physical, and mental teaching—the pupil is placed under entirely new and opposite conditions, by which bad thoughts are excluded, and good thoughts induced and impressed from day to day by practical work, from the like of which he may hereafter get something good for himself. In the morning he is roused out of his sleep to attend to his head-to-foot washing and his dressing. Then he has to go with others to his breakfast; after that to the school, where, with his class, he is kept to the simultaneous class lesson without waiting, to which he willingly gives himself, as it is not over-wearisome, like the lessons of the long-time schools. He may next have to fall under the drill-master or the gymnast, and, if he stumble or fail, he is jeered by the other pupils, or reproved by the corporal; but he soon participates in the zeal and competition of common lively action. He may on the following day have a swimming lesson. He may next have some naval exercise at the mast, where, unless he holds on, he will fall into the net spread beneath to receive him. Then he has to go to the workshop, where the work-master—in carpentry, in shoemaking, or in tailoring—keeps the mind, the hand, and the eye of the pupil intently occupied. His day's occupation may be varied by free-hand drawing, so useful to handicrafts, or by lessons in singing, or, if he be a very good and apt boy, by lessons in instrumental music. The enumeration of the incessant occupations may sound as of severe labour; but the course is varied by 'relief lessons,' and it becomes so little irksome that an interruption is disagreeable, and an exclusion from any part of it is acutely felt as a punishment. When some parents exercise their right of taking away children from the district school, the children are not glad, but commonly cry at having to leave the institution, to part with their playmates or their workmates, and to go home. As the physical and industrial exercises have been improved, desertions have diminished and the outcomes bettered. From morn until night bad thoughts are much excluded, and comparatively good thoughts—thoughts of doing better for themselves by work and wages, and by all honest and esteemed position—are generated and impressed. The teacher cannot look into the mind and see what effects, or whether any, have been produced by his precepts. But the drill-master or the work-master does see the valuable primary moral principles of attention, patience, self-restraint, prompt and exact obedience, in outward and visible action. The general result is that the pupil gets interested in what he does, and does it with a will.'

Reduction of Crime by Physical Training.—We are strongly of opinion that by the introduction of physical training the end will be accomplished of reducing natural crime. In the draft report before-mentioned, this important section is dealt with in the following terms:

'It is reported that the Vice-President of the Privy Council has acknowledged that the Code has not yet affected the state of crime. This acknowledgment may be cited as an instance of deplorable want of information of one department of what is going on in other educational departments, namely, in the district half-time schools under the Local Government Board, in the half-time schools under the War Department, and in the half-time school in action at Greenwich under the Admiralty, and the very effective work going on there. It would be seen that the Code has affected, and continues to affect, vice and crime, by creating inaptitudes to physical labour and productive service, with aptitudes to vice and crime, as displayed by the conditions described in the statement of Captain Brook, the manager of the very successful Industrial School at Feltham. But it has yet to be made known and understood how profoundly crime is affected by physiological conditions, which require to be dealt with by early training, including sanitation. This is shown, to a considerable extent, in the following answer of Dr. Guy, who for years superintended the Penitentiary, and had charge of the statistics of crime at the Home Department:—

'You are quite right in crediting me with the opinion that the physical condition of prisoners has more to do with criminality than the public are aware of. Of course this opinion results from my experience of convicts, and from returns relating to them. From censuses of the convict population, we have learnt that while nearly half of our male convicts are able-bodied, and more than a fourth healthy, but not robust, less than another fourth are ill, or fit only for light labour. Three in a hundred male convicts are of weak mind, insane, or epileptic. Eleven in a hundred are subject to scrofula and chronic diseases of lungs and heart. Twenty-three in a hundred have deformities or defects, congenital or acquired. The remaining sixty-three in the hundred have no deformities or defects. I think that these figures indicate inferior physical condition; but as we have no standard with which to compare them, the opinion must continue to be an opinion only. It is not a fact. I may add that among the three per cent. of convicts who are weak-minded, insane, or epileptic, are to be found those who have committed an undue proportion of the worst crimes, and also those who have committed the longest succession of petty offences. They form a formidable group of habitual offenders, for whom the proper place is the imbecile asylum. But the imbecile asylum, as distinct from the lunatic asylum, has still to be provided for country parts, and plans have to be devised for gathering our tramps and vagrants who belong to this class, and paupers too, into them. Our rural population, including the poor, properly so called, are greatly interested in this matter.

'Dr. Owen Rees, who for five years had medical charge of the Pentonville prison, states that his observations were all in accordance with the statement of Dr. Guy, which is also confirmed by the experience of Dr. Mouat of the Local Government Board, who has had the most extensive official observations of the condition of prison populations.

'The statement of Captain Rowland Brook, of the

Feltham Industrial School, as to the condition of physical inaptitude in which children are turned out who have been educated under Code, may be commended to particular attention. In a letter addressed by him to Mr. Chadwick, he concurs heartily and practically in the views now set forth :—" I quite agree with you that children will at all events make as much progress under a properly arranged half-time system as under the present long-time system; whilst, both politically and socially, the physical development of children, by technical instruction, should be insisted on in all the wage-earning classes. We have introduced gymnastic exercises here with marked success, the result being that boys, on leaving, are both mentally and physically prepared to earn their own livelihood. I am quite of opinion that the present Privy Council Examination much retards the progress of children in education by making the quick ones dance attendance upon the dull ones. Probably your *leaving examinations* would meet this difficulty. The present system of national education produces a dislike and inaptitude for labour amongst the children of the working classes, which, unless altered, will, I fear, produce very sad results. The sudden change from schoolroom to workshop is the cause of many children being unable to bear the trial they are called upon to undergo, and hence they become idle vagabonds and worse. All the boys admitted into this school are quite unable at first to work, and seem to have had no possible previous training to prepare them for it. It is only after they have for some time enjoyed the advantage of our half-time system that they acquire any willingness for labour." '

Promotion of Scientific Recreation.—Lastly, we submit

that, to ensure the happiness and serenity of the people of the future, the children of the present should have their mental and art training varied by making the subject of recreation a scientific branch of study amongst all who are engaged in educational work. In such advance we should have the means for recreation made the means for imperceptible instruction in bodily and mental powers, so that, having never unduly severed them from the tastes of the scholar, they shall be true resting-places, useful as well as pleasing diversions from mental and physical labour.

While this address has been in course of preparation I have received from Washington, from the Bureau of Education of the Department of the Interior, an article by Dr. Hiram Orcutt, on 'The Discipline of the School,' in which that eminent authority thus expresses himself.

'The mind and body,' he very truly observes, 'are inseparably connected. Hence mental culture cannot be successfully carried on without physical culture. Both mind and body must have recreation more than the ordinary recesses and holidays afford, and, as every teacher knows, there are certain hours and days when the fiend Disorder seems to reign in the schoolroom. He cannot assign any reason, but the very atmosphere is pregnant with anarchy and confusion. And what can the teacher do to overcome the evil? He may tighten his discipline, but that will not bind the volatile essence of confusion. He may ply the usual energies of his administration, but resistance is abnormal. He may flog, but every blow uncovers the needle points of fresh stings. He may protest and supplicate, scold and argue, inveigh and insist; the demon is not exorcised, nor even hit, but is only distributed through fifty fretty and fidgety forms. He will encounter the mischief successfully only when he encounters it indirectly. Here applies the proposed remedy—mental and physical recreation. Let an unexpected change divert the attention of the pupils.; let some general theme be introduced in a familiar lecture or exciting narrative; or, if nothing better is at hand, let all say the multiplication table or sing the Old Hundredth, and the work is accomplished. The room is ventilated of its restless contagion, and the furies are fled. Now add to this mental the physical recreation of school gymnastics, and the remedy is still more sure. Gymnastics are not only useful and important as a means of physical development, but also of school government. The exercise serves as a safety-valve to let off the excess of animal spirits, which frequently bring the pupil in collision with his master. It relieves the school of that morbid insensibility and careless indifference which so often result from the monotony and burdened atmosphere of the schoolroom. It sets up a standard of self-government and forms the habit of subjection to authority, and, as it is a regulator of the physical system, it becomes such to the conduct under the law. The gymnastic resembles the military drill, and has the same general influence upon the pupil that the military has upon the soldier, to produce system, good order, and obedience.

'Gymnastics also create self-reliance and available power. This is more important in life than brilliant talents or great learning. It is not the mere possession of physical power that gives ability, but the control of that power which this drill secures. Gymnastics preserve and restore health. It can be shown that the sanitary condition of schools and colleges has improved from 33 to 50 per cent. since the introduction of this systematic physical culture. Would we secure to future generations the realisation of the old motto, *Mens sana in corpore sano,* we must restore to our schools of every grade systematic physical training. True gymnastics are calculated to correct awkwardness of manner and to cultivate gracefulness of bearing. They give agility, strength, and ready control of the muscles, and thus tend to produce a natural and dignified carriage of the body and easy and graceful movement of the limbs.

'Again, the gymnastic drill awakens buoyancy of spirits and personal sympathy. Concert of action brings the class into personal contact in a variety of ways, and tends not only to create mutual good-will, but the greatest interest and enthusiasm. This promotes improved circulation, digestion, and respiration, and induces a feeling of cheerfulness and hopefulness that dispels despondency and every evil spirit. The gymnastic garb must leave the limbs free from restraint and the muscles and vital organs free from pressure. Hence, under this treatment, the beautiful form is left as God made it, to be developed according to His own plan. We must mark this as another advantage of gymnastics: to correct and control the ruinous habit of a fashionable female dress. Indeed, every department of education is carried on through a system of practical gymnastics. We have mental gymnastics, moral gymnastics, and physical gymnastics, which include vocal gymnastics.'

CONCLUSIONS.

I have now put forward our programme. It rests, as we conceive, on the bases of national necessities. A few concluding paragraphs may be taken as proposed resolutions to explain the mode in which we would carry out the reforms we have in view.

1. We propose to lessen the tasks of a mental kind in all schools by the general adoption of the half-time system, which Mr. Chadwick introduced into the factory over forty years ago. Believing that the brain of the child under fourteen years of age is sufficiently charged, to be safely charged, when it has been subjected to three hours' work in book teaching, we assume that this period per day of such teaching is sufficient for all useful and safe advancement, that the children would have as much as they could learn, and would retain more than they retain on the present plan. We propose at the same time to make inspection into book learning less critical and less severe, with the substitution of inspection into physical capability. In connection with this department we propose that there should be at stated times a physical inquiry, by competent authority, into the health of every school and every scholar, and that as much special encouragement and reward should be given to scholars who present the best physique as to those who present proofs of superior attainments in the standards. We propose further that this great change shall be effected by utilising the time taken from books and by applying it to lessons of play, exercise or work that shall be useful in developing the body, and in making it apt to attain proficiency in physical arts and sciences. We

would suggest that, in the school itself, the means for this physical instruction should be provided; but we would not by any hard-and-fast line hold by the school as the only place. If it were found in any case that a scholar had the means, in his half-time, of following any proper and profitable occupation without injury to himself, we should let him follow it, by which plan, as we believe, the sting of the compulsory clause in the Education Act would be most effectually blunted.

2. We propose that, while the mind of the child shall not be surcharged with book learning at a time when the body is in the most critical stage of development, either into a sound and helpful or into an unsound and helpless body, there shall be made a provision, in the school itself, by which education shall be allowed to go on after the usual prescribed time during which it is presumed that the education is completed. In this way, we believe the Voluntary and Board schools would become not only schools for the young but colleges for the old pupils, and, in respect to the old pupils, self-supporting centres of popular enlightenment.

3. We propose that the physical education introduced into our schools should be at once of the simplest and best kind, not a system of a particular character, but one which should combine everything that is useful in various systems, and which should interest the scholar while it develops his physical life. We agree with an observation made by Mr. Charles Roberts, in the letter from him already quoted, in which he says:—
'I have examined with some care, from a physiological point of view, the various systems of physical education, but I am not satisfied with any of them. The military

drill in use in many schools puts too great a strain on the lower limbs, and too little on the arms and trunk, and though the exercises are useful for discipline, they are monotonous and wearisome to children, and may be injurious, by inducing flat-foot and other deformities of the body. On the other hand, the exercises in ordinary German gymnasiums are generally too severe for children, and not sufficiently under the control of the non-medical teacher; their expense, moreover, places them beyond the reach of elementary schools. The Swedish system, again, as taught in the Board schools, lacks spirit and energy, from the entire absence of apparatus, and therefore of motive, to attempt or complete a definite object, a defect which Miss Chreiman's system has removed to considerable extent, by the limited use of simple apparatus.'

4. We propose that there should be introduced into the system what may be shortly explained as systematic training of the senses, so that the senses of the sight, hearing, touch, and even smell, should be brought up to the best standards of perfection. Such training, we are of opinion, could be carried out by means of lessons and use of simple apparatus, and would, in the course of carrying it out, afford facility for practically testing the capacity of every scholar, and his fitness or unfitness for the after duties he may be called upon to undertake. In America they have had appointed tests for the proof of colour-sight, so that it may be determined, when a man applies for duties in which colour-sights are required, whether he can distinguish colours. If our design were in operation, no scholar would leave a school without being made fully acquainted with his particular failure or capacity for any intended occupations.

5. We propose, finally, to use the time that we wish to extract from book-learning, in some, and, indeed, in a free degree, to the cultivation of certain of the more refined and pleasure-building arts. First amongst these we would place music as the primitive of recreative pleasures. We observe that our children are well and happy when they can sing; we see men and women gathered together, and find the height of mirth and happiness when somebody gives a song or tune. In the most refined society music is the joy of life; in the lowest dens men, hardly above animals, when they meet to be amused, sing. It may be that in all these positions the music is bad, but it is there, and the desire for it extends through creation. Here, therefore, is the first recreation to be scientifically studied. Make a nation, we say, a musical nation, and think how we have harmonised it socially, morally, healthfully. We cannot begin to teach this recreation too early or too soundly. We ought to begin by making the learning of notes in succession—the scale of musical chords—coincident with the learning of the alphabet. The one could be taught as easily as the other, and would be retained as readily, perhaps more agreeably. Next, the intervals should be taught, in a simple but careful way, so that melody may be acquired, and the art of sight-singing attained. From this elementary basis should follow the simplest forms of tune, after which a plain melody could be read with as much ease as the reading of the first story-book. Simple part-songs, leading to endless delight, would succeed, in exercise, and a true and natural language of sweet sounds would be the property, in one generation, of all the nation.

In addition to music we would, as a matter of course,

introduce other recreations, such as dancing, gymnastics, and all those muscular games and exercises which, by discharging, naturally, the nervous force, relieve the mind of mischievous intents and proclivities to destructive habits. This conclusion, we say again, is supported by experience. Visiting the schools at Anerley two years ago, we learned the fact, which has been so ably displayed by our learned American *confrère* above quoted, that the freedom of wholesome play and exercise is the safest of all provisions against what may be called lawless play or irresistible mischief. For purposes of economy, as it was supposed, a certain engaging and popular but rather expensive exercise was set aside there for a period because of the expense entailed. After a time it was discovered that various expensive mischiefs were being carried out, not in retaliation or from any desire to show an insubordinate spirit because the favourite game was withheld, but from actual inability to rest from the exercise which had previously been so healthy and agreeable. *Nolens volens* the nervous system discharged itself into the muscular, and in the process various substantial articles suffered. In the dormitories, bolstering became the vicarious amusement, and therewith the bedding underwent a destructive course which did not improve it for future wear. In the yards, ball-throwing and other like games assumed ascendency, by which windows were subjected to penalties which were more satisfactory to the glazier than they were to the management. Added to these unforeseen extravagances, personal exhibitions of prowess of an irregular kind became eminently fashionable pastimes, under and by which the clothing of the combatants was modified much more to the gratification of the tailor than the tailor's employers who had to settle with him. In brief, the expenditure incident to the proposed economy was soon found to exceed the expenditure for which the economy was instituted, and the result was a return to the lawful exercise, with every advantage to the children themselves, to those who were set over them, and to those who had to pay the expenses.

I have now submitted the programme which we who consider national necessities as the bases of national education would put before the nation in support of the grand revolution we suggest. Should it be urged that what we propose is too essentially physical or muscular, we answer that all education is, in the strictest sense, physical and muscular. Speech is muscular, expression is muscular, writing is muscular, composition is muscular, as much as mental. It is as purely a muscular act to decline a Greek verb as to walk across a tight-rope, except that the muscular movement, hardly so refined, is more concealed. We meet two men, one of whom is seen to move with ease and grace, the other with dulness and weight. We say, how accomplished the one, how uncouth the other! We hear two men discourse, the one with elegance, precision, style, the other with hesitation, blundering, rudeness. We say, how accomplished the one, how uncouth the other! In all these cases muscular force has played its equal part with mental aptitude or inaptitude. We see a man who has not been educated to grace of manner, or speech, or thought, assuming the part of a man of grace, manner, and thought, and, by much study, sustaining the character for a short time, as on the stage. But we know that the man only acts; he is not trained to the muscular skill that can carry him through all parts of life with equal grace, though he may, by intense labour, attain the minor part, and be perfect in it.

We know that no one who late in life enters a vocation requiring certain qualities, like that of a physician, a surgeon, a preacher, a pleader, a commander, a pilot, an engineer, a player, can gain that full self-possession which comes, as it is said, naturally, to the man who has been from early life trained in the work. Here, again, the failure we affirm is muscular as much as mental. The concealed muscular mechanism is not in working order. The mind may issue its commands, but, if the muscles fail to obey, the mind, like a general whose redcoats are undrilled and impervious, may break itself into imbecility and produce no results beyond hopeless and helpless confusion and defeat.

So we contend for the physical education of all our young, on the lines I have laid down, as the stirring want in this stirring time. Our intention is to make this nation a nation of heroes as well as scholars; a nation that the sculptor can describe as well as the historian; a nation that can hold its own in the scale of vitality, and protect its own by the virtues of courage, physical prowess, and endurance, as ably as by statesmanship and knowledge, more ably than by expediency and craft. In all which efforts we accept and act on the motto which our leader has riveted on our standard—

Primo vivere, deinde philosophari.

DISEASES INCIDENT TO PUBLIC LIFE.

POLITICAL Niagara now and again coming into full play, with advents of new Parliaments refounded on new lines, is ever giving impulse to all sorts and conditions of men to enter the political arena and seek in new scenes of strife for the bubble reputation.

Whether this political torrent is to be for good or bad for the country at large, whether it is going to breed statesmen or underbreed statesmanship, are points on which a physician need offer no opinion. For all that, he has a duty which may be conscientiously discharged, which is entirely within his own province of observation, and which, touching as it does the fate of new national experiments in certain of their stages, is in correct place in a volume on the common health.

To all who have had under their observation the phenomena of motion in the physical and mental life of man until the phenomena have become an open book, it is clear that coming political events will, if they do nothing else, lead to great changes in the health of those who, as legislators, are to play the most important parts, for good or for evil, in the impending revolution. In plain words, some of the men who may succeed in entering the House will, of necessity, suffer from diseases which are almost inseparable from public life, and the risks of which they must meet in accepting the so-called honours which their ambition or their desire to serve their country leads them to obtain.

The character of the diseases that will be manifested will vary with the characteristics of the individual for the nature of the work that has to be done in the parliamentary struggle. In this work there are three perils, two of which affect different persons differently, and one of which affects all more or less, as a common peril. The first is the conflict, always in progress, connected with hopes deferred, passions ignited, excitements of victory, depressions of defeat. The second is the outdoor work of the parliamentarian; the preparation for debate or vote; the perusal of the public opinion expressed outside the House by speeches, leading articles, and correspondence, and the friction of mind and physical labour connected with attendance at public meetings, committees, deputations, dinners, and the various other public entertainments. The third peril which affects all in common is the late sitting, the night work of the legislating labourer, with, on latest nights, the last great excitement of the conclusion of debate at the time of year and at the hours when, under favourable conditions, the human body is at its lowest natural ebb; the hours of hours in which the stricken to death most frequently yield up their last breath to the eternal space.

The diseases most commonly incident to the public life are almost entirely those which affect the bodily organisation through the mental surface.

They are the results, either of intense shocks driven

through the senses from without and deranging the molecular or molar construction of surface; or of persistent friction between the immutable spirit flowing in regular tides into the mutable flesh, and preventing its repose for efficient repair of its structure; or persistent friction on one particular centre of the mental surface, with failure of it and of all parts subordinate to it as a centre of power.

Thus in the phenomena of disease which we observe in the men unfitted constitutionally for public life, the grand series are diseases of that fleshy physical surface, the nervous system, on which the mind is distributed; that resident medium of the mental motion derived from outside the organism, which, entering the *camera nervosa* by the senses, is reflected back in act, word, deed.

The diseases incident to the series of induced changes by mental shocks are most developed in emotional and animal emotional men. The shocks tell in them, through the sympathetic nervous organisation, on vital organs, and especially on the heart. The heart is being kept in continual perturbation, its nervous supply irregular, intermittent, palpitating. Whatever other part of the body may be deprived of power, the regulating heart must never be deprived. 'First come first served' is the rule of the heart; and so it takes, literally, for its own use, and for its own special nourishment, the very first drops of blood which it pumps out at each of its strokes. If it fail in this process, through the wrong done to it by its owner, it immediately suffers. The coronary arteries—the primary branches of the great arterial tree which rises from the left ventricle of the heart—failing to bring round to the muscular walls of the heart sufficient food for their support, the nutrition of the heart itself is earliest to suffer; then, as the nervous supply which prompts the muscular action is reduced, there occurs a further failure; and, lastly, as the nervous surface which is the seat of the nervous power must have its required blood supply reduced when the centre fails, there is a third bad effect telling back upon the central vital organ. The whole leads to a disablement from the centre under which many public men succumb during subjection to excitements they were not able to encounter.

The illness and death of Mirabeau, as told by his compatriot Cabanis, is a perfect illustration of an emotional statesman killed from the heart by excitement, killed as if he had been stabbed to the heart and had died a lingering death from loss of blood.

In one of our public institutions there is a model likeness of one of the political heroes of the working-men of this country, who died so like Mirabeau that if Cabanis' account of the death of the great statesman were changed by the mere alteration of the names of the sufferers, the clinical record would read for either the one or the other. They were, socially, very different men—the one a patrician and scholar, the other a plebeian and a self half-taught. But constitutionally they were the same, in accident of political strife the same, in result of disease the same, in suffering the same, in mode of death the same. At the generous instance of the late Mr. Danby Seymour, I was the consultant called to render physician's aid to the English political enthusiast, who was as sincere a man as he was intense and unselfish. And my return of him in the clinical note-book reads, 'Intermittent circulation and disorganised heart from political excitement and enthusiastic worry.' This man told me that he dated the knowledge of his symptoms, his breathlessness and failure of central power, to some rough physical usage to which he was subjected while descending from a waggon from which he had been addressing a multitude. I have no doubt that was the turning-point of his failure, but it had been led up to by long previous strain.

A few professional and political friends, including, on the professional side, the late Dr. Edwin Lankester, and myself, were steaming down the Thames to see the great outfall for the London sewage. We had with us a friend we knew well, as a scholar, a well-known publicist, and a vehement emotional politician, with an historical name as such. On the way this enthusiast broke forth into one of his violent political arguments, and while speaking suddenly stopped. He held out his hand for me to feel his pulse, which was intermitting as if it would cease altogether. When he had quite calmed down he took me aside, and described to me his condition, the same, in kind, as told by Cabanis of Mirabeau, but less advanced. I warned him: 'If you repeat these vehement outbursts, you will one day empty your heart into your brain, and then and there end all your fervour.' He called out to Lankester to come, and repeated to him my statement as a saying, pithy, he thought, but rather a play upon words than of serious import. 'I agree,' said Lankester, 'and would advise you to take the hint.' He, however, the man most concerned, treated the matter as a false alarm, went on in his career unchecked, and one day, not long afterwards, while delivering in public an impassioned speech against American slavery, died on the platform, as we had foretold. He emptied his feeble heart into his powerful brain. These are extreme examples in respect to the sudden endings of public life, but are of great value as lessons. They are the understandable illustrations of a number of other examples, minor in appearance as phenomena, but only just short of seriousness in fact. In the prolonged form in slow collapse from central failure, in breaking down, as it is commonly called, such examples are numerous amongst public men of the emotional type, and if the experience of other physicians be the same as my own, they are, year by year, on the increase.

Sometimes, especially amongst men of the emotional, and still more amongst men of the animal emotional type, an opposite condition obtains. In their excitement, commenced early in life, their heart becomes too powerful for their body. It becomes an uncontrollable organ. Its nervous organisation is most impressionable, its own brute power, if I may use the expression, out of balance with the rest of the body. In these cases the mental nervous surface is the sufferer. The man is said to be very excitable, to fire off quickly, to be unmanageable by any authority. For a time this power is long evidenced, and troublesome to all whom it affects; then it is noticeable as being subject to sudden arrest in him in whom it is manifested. He is violent as ever for a period, but on slight opposition he sits quietly down, reposes a while, and soon gets up again, repeating the act until the repetition becomes a more serious impediment to business than the former long and vehement harangues. Lookers-on say of the man, 'Let him have it out, and have done with it;' they try the experiment, but it does not succeed, for he never has done with it. He has got into the fits and starts stage of emotional disease. He is the irresponsible possessor of a strong unbalanced heart hammering a feeble brain into illucid intervals of helpless imbecility. Men of this kind sink into disease from the brain. They often die of apoplexy, which in plain English means a blow or stun, as if they were knocked down like oxen by a blow on the head, as in fact they are, except that the last blow is delivered from the inside instead of the outside of the organism.

In other examples the blow is not sufficient to kill outright; it is a 'stroke,' a stroke of palsy—so it is expressed. One controlling part, and no more, of the nervous surface gives way; the arm falls, or the lip falls, or the half of the body loses its power, and the correct rendering of words ceases. There is paralysis and aphasia.

In a third class of this same type the process of change is more gradual. The fire burns lower, but does not die out. The man, advised by his physician and his friends, wisely retires from combat; he is too easily excited to bear it. Occasionally such a man finds out his peculiar constitution in time to save himself from immediate danger. He retires early from the public arena into a calmer sphere, and, by sufficient watchfulness, lives on for many and often useful years. As I write I recall an instance of this kind in a member of my own profession, who discovered his danger early, gave up all thought of public conflict, and retired into a quiet medical practice, in which he was both active and successful. By this plan he lived on for over thirty years. Death came, however, at last suddenly, according to the rule so common in such examples.

In another set of these public men the saving of life is achieved to an extent, but not effectively. Life is prolonged, but healthy life is lost. The *vis nervosa* is no more. The need for change, for repose, for freedom from responsibility, is ever the necessity. The life is a penance for ambition, and is always short.

There are illustrations of another set of phenomena of disease incident to public men, in which the vital failure, springing originally from nervous injury, takes the course of a purely physical wasting malady. One disease of this character stands pre-eminently forward. I mean diabetes. I have myself had under my medical care at least six public men in whom this disease was clearly developed by some mental perturbation or nervous shock. In one of these instances the sufferer, for a short time a well-known and important officer of the State, had come into office at a period of extreme political strain. Solemnest issues were at stake, and momentous duties and decisions rested upon him, a man, of moderate capacity, who, by pure industry and perseverance in a laborious profession, had won an esteemed position, and who, just suiting the master then at the head of affairs, got into an office which he himself said was heavier than he could bear, and yet could not, with honour, leave. Called to question in the House on one of the responsibilities of the century, this statesman had a sense of faintness while speaking, and from that moment, under the mental strain or shock to which he was subjected, became diabetic, probably from some mechanical accident in his brain, involving the parts about the fourth ventricle. Thenceforward he died as straight off as a man can die from diabetic affection in its steadily progressive form.

The peril which is incurred by all political men in this country when they get a seat in the Commons, the peril of broken rest, has its forthcomings. Strong or weak, the members feel this risk, but not all equally. Some who are sufficiently fortunate to be able to turn day into night and night into day, and who are endowed with the faculty of sleeping at any time, get on very well, by their incessant labour astonish the world, and, unfortunately, often mislead their feebler brethren into attempts of a similar kind. But these are quite exceptional men. The majority, who are unable to give time for sleep in the day, sink into sleepless decrepitude; feel, at last, that they cannot sleep when the opportunity offers, and, depending on narcotics for a false repose, become, by a second nature, the slaves of a habit which has but one termination.

To the list of diseases above rendered as incident to active public life might be added those affections which we physicians call diathetic, and which are hereditary in those who suffer. It is possible that some diathetic diseases are brought out by the wear and tear of public work, but, then, they are also easily brought out by other and opposite influences, like gout when it is excited by luxury and too much rest. Setting one thing against another, hereditary tendencies, with one grand exception, need not trouble the would-be legislator.

But there is the one exception, which should never be left out of the reckoning, and that is insanity. The man who knows himself to be afflicted with that taint is doing his first insane act, while yet sane, when he throws himself into the struggles of a political career. This is no paradox; it is a truth which experience of men and of the proclivities of men teaches with a repetition that admits of no denial.

To one standing aside and looking on at the political turmoil, it is marvellous that so much risk of life, vehemence of opinion, and utter uselessness of contention and labour should agitate the world. It can only be that political work is utterly wanting in all scientific accuracy that so many men so madly rush into it. Half a dozen great political scholars, working half a dozen months on a text-book of political science for the use of schools, might do more for the world, if their labours were accepted, than half a dozen parliaments, sitting out their full terms, and fighting each other from the beginning to the end of the chapter. Such fighting and killing, however, must be, until the long day when the truth dawns, that in matters where every one thinks he knows more than every one else, only a few know anything. But, while that day of light is dawning, let no one rashly believe that, in spite of all constitutional weaknesses, he can go on the political war-path with safety, either to his physical or to his mental organisation.

WOMAN AS A SANITARY REFORMER.[1]

Two of the wisest of men, and by necessity, therefore, both of them Sanitarians, Solomon and Xenophon, have laid down rules bearing on the duties of women who rejoice in being called wives as well as women. 'A good wife,' says Solomon, 'worketh willingly with her hands.' 'She is like the merchants' ships, she bringeth food from afar.' She is an early riser, and sees that every one has an early breakfast. 'She riseth while it is yet night, and giveth meat to her household and a portion to her maidens.' By exercise she strengthens her limbs. 'She layeth her hands to the spindle and her hands hold the distaff.' She knows that where there is poverty there can be neither health nor happiness. 'She stretcheth out her hands to the poor; yea, she reacheth forth her hand to the needy.' She provides against the cold. 'She is not afraid of the snow for her household; for all her household are clothed in scarlet.' In clothing herself she combines artistic taste with usefulness, as every woman is bound to do. 'She maketh herself coverings of tapestry; her clothing is silk and purple.' 'She maketh also fine linen and selleth it.' 'Strength and honour are her clothing.' She combines common sense with gentleness. 'She openeth her mouth with wisdom;

---

[1] Lecture delivered to the Congress of the Sanitary Institute at Exeter on Thursday, September 23, 1880.

and in her tongue is the law of kindness.' She is watchful and busy. ' She looketh well to the ways of her household, and eateth not the bread of idleness.'

And these, says this wise Sanitarian, are her rewards: 'She shall rejoice in time to come.' ' The heart of her husband doth safely trust in her.' And, light of perfected human happiness! ' Her children rise up and call her blessed.'

The second of the wise Sanitarians, Xenophon, tells his story of the good wife in somewhat different terms and manner, and indeed with difference also of detail. He, treating of the household and of the economics of it, invents a dialogue. He makes Socrates and Critobulus hold a discussion which comes to this general understanding: that the ordering of a household is the name of a Science, and that the Science becomes the order and the increase of the house. Afterwards, Critobulus asks of the Master why some so use and apply husbandry that they have by it plenty and all the good things that they need, including health and all blessings; while others so order themselves that every good thing avails them nothing at all. ' These two things,' says Critobulus, ' would I like to have explained by you, to the intent that I may do that which is good and eschew that which is harmful. Thereupon, Socrates, the Master, recounts to his pupil that he once held a communication with a man who indeed might be called a good and honest man. He had already seen and studied the works of good carpenters, good joiners, good painters, good sculptors, and had seen how they attained to excellence; and so he desired to find out how they who had repute for goodness and honour attained their excellency. He looked for such an one first amongst those who were handsome, but it would not do; for he found that many who had goodly bodies and fair visages had ungracious souls. Then he bethought him to look for a man who by general sentiment was reckoned upon as good, and at last he found Ischomachus, who was generally, both of man and of woman, of citizen and of stranger, called 'the good.'

Socrates is made to discover Ischomachus sitting in the porch of a temple, and, discussing with him many subjects, asks him how it is he is called a good and honest man. At this Ischomachus laughs. 'Why,' he replies, 'I am called good when you and others speak of me I cannot say. I only know that when I am required to pay money for taxes, priests, or subsidies, they call me Ischomachus; and indeed, Socrates, I do not always bide in my house, for my wife can order well enough whatever is wanted there.' 'And did you yourself bring your wife to this perfection,' asks Socrates, 'or did her father and mother teach her?' 'As she was but fifteen when I married her,' returns Ischomachus, 'she had seen very little, heard very little, and spoken very little of the world; and therefore'—he continues some way further on—'I questioned and then instructed her.'

It is very doubtful whether, in these days of supreme wisdom, the first principles of Ischomachus, as he taught them to his beloved, would be at all permitted. I dare not certainly set them forth on my own account, although they bear directly upon the subject of my lecture. I record them, consequently, as I find them, leaving their author, Master Xenophon, who though dead yet speaketh, to assume the responsibility of so flagrant a series of propositions as will follow.

'Methinks, then,' says Ischomachus, 'that for the welfare of every household there are things that must be done abroad, and things that must be done within the house, and that require care and discipline.' We shall probably be all of one mind, even now, on that point. The difference of opinion that will rage rests on the succeeding points of argument. 'Methinks also,' he continues, 'that the God hath caused nature to show plainly that a woman is born to take heed of all such things as should be done at home, and these are the reasons for the belief. He, the Maker, hath made man of body, heart, and stomach, strong and mighty to suffer and endure heat and cold, or privation, to journey, and to go to the wars. Wherefore, he hath, in a manner, charged and commanded him with those things that be done abroad and not of the house. He, also, remembering that he has ordained the woman to bring up young children, has made her far more tender in love towards her children than the man. And, whereas he has ordained that the woman should keep those things which the man getteth and bringeth home to her, and knowing also that to keep a thing safely it is not the worst point to be doubtful and fearful, he has dealt to her a great deal more fear than he did to man, while to man, who must defend himself and his own, he has dealt out more boldness. But because it behoveth that both man and woman should alike give and receive, he has bestowed on them alike remembrance and diligence, so that it is hard to discern which of them has most of those qualities. He has, moreover, granted them, indifferently, the power to refrain from doing that which is wrong, so that whatever either of them does better than the other is best for both ; and because the natures and dispositions of them both are not equally perfect in all these things, they have so much the more need the one of the other; for that that the one lacketh the other hath. Likewise the law shows, and the God commands, that it is best for each to do their part. It is more correct for a woman to keep house than to walk abroad; and it is more shame for a man to remain skulking at home than to apply his mind to such things as must be done abroad.'

Ischomachus is next shown as teaching his wife the lessons to be learned from bees in the hive; and, improving the occasion, he offers certain rules which, with actual reverence, I venture to epitomise. He was, in fact, a grand Sanitarian, this Ischomachus, and I do not think there would have been much sickness now in the world had wives, in general, been after the training of Madam his wife. The million of men of physic might have become reduced to thousands. And the women of physic? Well, I may relieve my mind at once and say it plainly; they might have become housewives of all houses.

Some lessons of economy are first to hand. The wife is to beware that that which should be spent in a twelvemonth be not spent in a month. The wool that is to be brought in is to be carded and spun that cloth be made of it; and the corn that is brought in must be most carefully examined, that none which is musty and dirty be eaten as food. Above all, the same instruction that Solomon insists on is enforced with special fervour. The wife is to be most particular, if any of the servants fall sick, that she endeavour herself to do the best she can, not only to cherish them, but also to assist them that they may have their health restored.

A little further on the philosopher touches on the importance of perfect order in the house as connected with the health and wealth of it. He tells how he once went on board a ship of Phœnicia, and wondered that in so small a space such a vast number of things could be stowed away with so much neatness that everything could be found in a moment, even during the confusion of a storm.

From these lessons he teaches his wife, and thereby all wives, matters that are more particularly of a sanitary kind. A house, he says, has an ordination. It is not ordained to be gorgeously painted with divers fair pictures, though these may be excellent, but it is built for this purpose and consideration, that it should be profitable and adaptable for those things that are in it, so that, as it were, it bids the owners to lay up everything that is in it in such place as is most meet for the things to be put. Therewith he disposeth of places for things in due form, and assigns the uses of the various parts of the establishment, in such manner that the woman who presides over the whole shall know the parts in a truly scientific way.

The inner chamber or room, because it stands strongest of all, is to be the strong-room in which the jewels, plate, and every precious thing in the belongings of the house must be securely located. The driest places are to be places for wheat; the highest places for such works and things as require light. The parlours and dining places, well trimmed and dressed, are to be cool in summer and in winter warm. The situation of the house is to be towards the south, so that in winter the sun's light may fall favourably upon it, and in summer it may be in goodly shadow. The wearing apparel is to be divided into that intended for daily use and that required for special or grand occasions. Everything belonging to separate service, to the kitchen, the bakehouse, the bathroom, is to be assigned to its own place and use. All instruments which the servants use daily are to be shown to the servants in their right places, and are to be kept there when they are not wanted. Such things as should not be made use of except on holy-days and rare occasions are to be left in special charge of an upper servant, who should be instructed beyond the rest of the servants to observe the same rules as the mistress herself would carry out. 'At last, good Socrates,' said Ischomachus, 'I did express to my wife that all these rules availed nothing unless she took diligent heed that everything might remain in perfect order. I taught her how in commonwealths, and in cities that were well ruled and ordered, it was not enough for the dwellers and citizens there to have good laws made for them unless they chose men to have the oversight of those laws. In like manner, then, the woman should be, as it were, the overseer of the laws of the house, as the Senate and the Council of Athens oversee and make proof of the men of arms.'

Finally Ischomachus touches on the mode by which his wife should maintain her own health. He observed about her, as a very strange habit, that upon a time she had painted her face with a certain unguent that she might seem whiter than she was, and with another unguent that she might seem redder than she was; and that she had a pair of high shoes on her feet to make her seem taller than she was. Whereupon, 'tell me, good wife,' said he, 'whether you would judge me worthier or more beloved if I explained to you what we are precisely worth, keeping nothing secret from you, or if I deceived you by declaring I had more than I really had, showing you false money, chains of brass instead of gold, counterfeit precious stones, red instead of scarlet, and false purple instead of pure and good.' She replies, 'The gods forbid that you should be such an one.' He then recalled to her her own deceptions; and when she inquired how she should be fairer in reality and not appear so only, he gave her as counsel that she should not sit still like a slave or a bondwoman, but go about the house like a mistress and see how the works of the house go forward, look after all the workers and sometimes work with her own hands, by which exercise she would have a better appetite for food, better health, and better favoured colour of her face. Then, likewise, the sight of the mistress, more cleanlily and far better apparelled, setting her hand to work and, as it were, striving at times with her servants who should do most, would be a great comfort to them by leading them to do their work with a good will instead of doing it against their will. For they that always stand still like queens in their majesty will be only judged of by those women who are triumphantly arrayed. 'And now, good Socrates,' continued Ischomachus, in conclusion, 'be you sure that my wife lives even as I have taught her and as I have told to you.'

Were a modern sanitary scholar, such a one as now speaks to you for example, to presume to lay the basis of sanitary reform, through the influence of woman, on such simple rules as those given above, he might suffer for a trouble, which would, in truth, be called a presumption.

Happy, therefore, is it that he finds the basis ready laid by two such masters as Solomon and Xenophon. Their sufferings are over, hidden in the inaccessibility of historical distance. Their words alone remain faithful as ever, and as true for to-day and for to-morrow as on the days when they first went forth. They are the basis of modern sanitary law with women as its administrators. I would not dare to add a syllable to their majestic common-sense. Good wives of the type of the wife of Ischomachus would, in one decade, make domestic sanitation the useful fashion and order of the nation they purified, beautified, and beatified.

I quote this basis of wifely work and duty because I feel more deeply, day by day, that until it is admitted, and something more built upon it, sanitary progress is a mere conceit, a word and a theory, instead of a thing and a practice. It is in those million centres we call the home that sanitary science must have its true birth. It is from those centres the river of health must rise. We men may hold our Congresses year after year, decade after decade; we may establish our schools; we may whip on our lawgivers to action of certain kinds; we may be ever so earnest, ever so persistent, ever so clever; but we shall never move a step in a profitable direction until we carry the women with us, heart and soul. Adam had no paradise in Paradise itself until Eve became the helpmeet for him. How then, in a world which is anything but a paradise, shall we transform it into anything like one till the Eves lend us a hand, and, combining their invincible power with ours, give us the help that is essential to success? We must go entirely with Xenophon in the belief that the human being is not perfected, either in thought or action, until the two natures are blended in thought and action. The man invents, the woman applies the invention; the man conquers nature, the woman makes useful the victory; the man discovers, the woman turns the discovery to due and faithful account; the man goes forth to labour, the woman stays at home to watch the centre common to them, and tend the helpless there. Yet both have remembrance, both have diligence, both have the power to refrain from doing what is wrong, and whatever either of them does better than the other is best for both. And, because the natures and dispositions of both are not equally perfect, they have so much the more need the one of the other, since what one lacketh the other hath. In the art of cultivating Sanitary Science this mutual understanding is necessity itself.

We ought not to blame womankind because it seems that women are behindhand in the work. They are not, in point of fact, behindhand at all; they are rather the forerunners in the race. Long before the word Sanitation was heard of, or any other word that conveyed the idea of a science of health, the good cleanly thrifty housewife was a practical sanitary reformer. Nay, if we come to the question of organisation itself, we have in this country, in that admirable institution the Ladies' Sanitary Association, the first of the great sanitary societies which, by its publications, its practical aid to mothers, its outdoor recreative parties to the stived-up children of the metropolis, and by various other means, has set an example that will one day be historical as a part of the great movement in the promotion of which we are engaged.

There is not, therefore, one single difficulty in the way of making the woman the active domestic health-reformer. The only thing that requires to be put forward is the method of bringing her universally into the work, and, if I may so express it, making the work a permanent custom or fashion, to neglect which would be considered a moral defect. There are in England and Wales alone six millions to be influenced. The first suggestion is that the beginning of the crusade shall be a beginning that shall not drive, but lead; that shall not dictate, but patiently suggest.

If what Pope said of man be true—

Men should be taught as though you taught them not,
And things unknown be told as things forgot,

in respect to the sex still more susceptible and impressionable, especially when those truly feminine duties which are connected with domestic health and happiness form the subject of advancement, it may with equal truth be said—

Women should ne'er be taught a thing unknown,
It should be credited as all their own.

Nor can any finer or nobler occupation be imagined than is implied under this head of domestic care and nourishment of health. There are women who think it the height of human ambition to be considered curers of human maladies; content at best to take their place with the rank and file of the army of medicine, and not perceiving that the only feature in their career is its singularity, a feature that would itself become lost if the wish of the few became the will of the many. I would not presume to interfere, on this point, even with the wish of the few. At the same time I would with all my strength suggest to women that, to become practitioners in the *preventive art of medicine*; to hold in their hands the key to health; to stand at the thresholds of their homes and say to disease, ' Into this place you shall not come, it is not fitted to receive you, it is free only to health, it is a barrier to disease;' to conjoin in this work so effectually as to be able to say to every curative practitioner who invades their cities, ' You may come if you please, and settle down if you please, but there will be nothing for you to do, except to write up, after a time, as a warning to all practitioners of the curative school: " Who enters here leaves hope behind ;"—to exercise practical power in such a manner would, I venture to indicate, be as much above the exercise of curative art, as the art of making unsinkable ships would be above the toil of working at the pumps of a sinking vessel, that was only sinking because it had been allowed to fall into a perfectly hopeless state and condition for resisting the strain of the deceitful sea.

I press this office for the prevention of disease on womankind, not simply because they can carry it out; not simply because it pertains to what Xenophon describes as their special attributes, their watchfulness and their love; but because it is an office which man never can carry out; and because the whole work of prevention waits and waits until the woman takes it up and makes it hers. The man is abroad, the disease threatens the home, and the woman is at the threatened spot. Who is to stop disease at the door, the man or the woman? What does a man know about a house, about the very house he lives in, if he be a man employed at all? I asked as good a man of business as ever went on 'Change how many rooms he had in his house. His reply was, 'What an absurd question!' 'Why absurd? the house is your own.' 'Yes, but I have never thought about it. You should ask my wife if you want to know. She will tell you all about it from the butler's pantry to the cockloft; but as I only go into two or three rooms myself, how should I be likely to remember? It is not my department.' That is so generally. The woman knows all about it, or if she does not she ought; it is in her department to know the whole matter by heart. The house is her citadel.

There probably is not a person who is given to reflect who will not in the main agree with me in these conclusions. The strongest-minded woman, the woman who would assert to her heart's content the right of womanhood to assume manhood, would, I think, agree with me in the main. She might and possibly would affirm that I do not go far enough; she might feel the position I have assigned to woman as too feminine in its tenderness, and as a retrogression from the design of attaining the equality of power which she would consider necessary for the perfect liberty of woman from the bondage imposed by men. At the same time she would agree so far as to admit that, if her fellow-sisters everywhere could claim and hold and maintain such a power of practical knowledge and skill as I have pointed out, their mission in this world would be more greatly advanced and more nobly utilised than it is at this time. Nay, perchance when she has heard me to the end, and has well considered the tremendous power which a completed scheme would give to her sex, she might feel that her ambition would be more than satisfied by its accomplishment.

While women in general will, I feel sure, almost think it impossible that so much useful influence could be attainable, the majority will ask, ' By what process of training can we so govern domestic life that diseases may be prevented wholesale; that life in all its innocence and fascination may never, except by the most vulgar accident, be invaded by death; that adolescence in all its beauty and unfolding strength may be equally guarded; that manhood and womanhood may have the same protection; that middle age may be extended in intellectual and physical perfection into the grand decline; and that the grand decline itself may be so gentle, so peaceful, so beautiful—yes, so beautiful, for there is a beauty in healthful old age that is unsurpassed—that life shall be but a dream, and death but a natural sleep? They will ask, I repeat, the majority of them, by what process of training can we help towards a triumph of science so beneficent?

I devote myself from this point of my discourse to give some answer to that question. I state at once that the training required is simple, beyond simple; that every woman who wills to go through it may go through it and may become mistress by it of the destinies of the world. Not the Fates themselves were more the mistresses of the destinies of the race than the women of an educated Commonwealth, who were fully conversant with the art of the prevention of disease and premature decay.

Ischomachus, content to have his wife taught housewifery pure and simple, would, I think, in this day be not quite so content. He would wish that she should know everything about the house in which she and he and their family dwelt; he would wish also that she should know something of that house of life which belongs to herself and to all hers. He would not desire that she should become a profound anatomist; he would not care for her to enter on the subject of experimental and practical physiology; he would scarcely aspire that she should try to emulate Hippocrates in diagnosis, or Dioscorides in therapeutics. But with our modern knowledge in his possession he would, I venture to suggest, have begged of her to learn a few principles which would help her to understand the reasons for the necessity of her domestic cleanliness and wifely care. As he is gone before these desires could be current, I will, with much respect, take his place, and indicate what every woman who aspires to be a wife, a mother, and a practical Sanitarian ought to learn in this particular direction.

She should master physiology so far as to understand the general construction of the human body. She should know the nine great systems of the body: the digestive, the circulatory, the respiratory, the nervous, the sensory, the absorbent and glandular, the muscular, the osseous or bony, and the membranous. She should be led to comprehend the leading facts bearing on the anatomy and function of these systems. She should understand what part food plays in the economy; the nature of the digestive ferments; the primary and secondary digestions; the method by which the digested aliment finds its way into the blood; and the specific purpose which is answered by and through the application of foods, proximate and elemental. She should be rendered fully conversant with the different changes of food that are required for the digestive process in different periods of life; the extent to which the digestive powers should be taxed in infancy, childhood, adolescence, maturity, first and second decline, and old age. She should be made aware what substances, taken as food, are of real and what of spurious quality. She should be taught the relationship which solid foods hold to liquid foods or drinks. She should be told what drinks are foods, and she should specially understand what are the particular foods required for the young during the periods of active growth. In illustration of the value of this last-named fact, it may be stated that if women only knew what foods were requisite to feed the skeleton or bony framework of the living body while that skeleton is in the course of growth, and if she would act upon her knowledge, as she almost certainly would if she possessed it, there would hardly be one deformed child left in the land in one or two generations. Rickets, with all its attendant miseries of bowed legs, crooked spines, and humped backs, would pass away as if by the spell of an invisible enchantress.

After the understanding of the digestive system, the woman should learn the principal facts relating to the circulation of the blood, the organs of the circulation, the heart, the arteries, the capillaries, the veins and the blood itself. She should know completely the mechanical construction of the heart, its coverings, its cavities, its lining, its valves, and the uses of the parts. She should understand the work of the heart; how it rests when the body reclines; how easily its daily tonnage of work can be increased by perfectly unnecessary strains and stimulation until a day and a quarter of hard work may be compressed into one day, and a fourth of the vital spring of the heart for that day be lost for ever, as so much taken from the sum total of life. She should know how the heart is sympathetically moved in its action, and may be weak or strong, regular or irregular, calm or excitable, by the influence of external impressions which, in passing, may seem nothing and yet be everything. She should learn that in early days the whole after-life may be shaped, I may say, by the tone that is given to the heart, and that whether in its pilgrimage a Faintheart or a Greatheart shall occupy the stage on which a young life is to enter shall turn absolutely on this one educational fact, the skill of the trainer of that simple and susceptible mechanism, the human heart, while yet it is susceptible, fashionable, and undetermined.

Nor should she, in regard to the healthy organism, be less informed respecting that breath of life which is ever being breathed into the living being by the Eternal Chemist, whose constructions and resolutions are the motions visible and invisible of his eternal universe. The complete structure of those breathing lungs should be as plain before her as the outward form of the things she knows best. The course of the blood, like a curve from one side of the heart to its other side through the maze of spongy lung-tissue, should be easily traced. The expansion of the six hundred millions of little vesicles of the lungs, which the air inflates, that it may, over so vast a surface, expose itself to the circulatory blood in its rapid passage over the vesicles; the change that takes place in the blood during the passage; the gas that is robbed from the air by the lungs; the gas that is given up to the air by the lungs; the change in the colour and character of the blood that attends these processes; the course of the changed blood bearing its vital air or oxygen, in myriads of tiny cells, through the arteries to the body at large; the spreading out of this blood over the vast sheet of minute vessels which make up the vital expanse, the vital furnace, the vital foundry of the body; the consumption of vital air there; the unloading of new material or pabulum there; the removal of old and effete structure there; and the recharge of the blood with the gaseous products of animal combustion there:—these things ought to be as familiar to the mind of our scholar as the commonest things in life: the letting in of air to feed the fire, the entrance of the servant with coals, the burning of the fire in the grate, the use of the fire for various domestic purposes, the opening of the ventilator to allow the smoke to ascend the chimney, and the removal of the ashes and *débris* that more new fuel may be supplied to keep the fire alive.

Equally clear to her should be the leading facts bearing on that receptive system of the body into which the external universe transports itself, and from which, in reflex response, the acts of life, the expressions, the movements, the thoughts, return in wavelike repetition back again, to become themselves external phenomena, linked, as such, with all the visible universe. Those nervous centres, locked up in the skull and spinal cord, to receive and retain and remit; those doubly-acting nervous cords, bearing the impressions to the centres and bringing them back again; those exquisite nerves, so finely set and balanced and distributed for play of reason and volition; and those sympathetic, nervous centres in the trunk of the body, allied to the viscera, which they serve, and governing the automatic motions on which the volition has but indirect control, centres of emotional and what is commonly understood as instinctive faculties;—these parts, these systems, in respect to general function and vital value, should all be as familiar as the course of the sun, from whom, in essence, they spring. And with these nervous organisms the fields of the senses, too, should be made clear; the outlines of the plan of an organ of sense being as simply comprehended as the plan of a camera or other well-known human instrument.

Let me interpose one practical illustration here of the value of knowledge bearing on the organs of the senses. Recall how many young people and middle-aged people are going about in spectacles, unable to see any object with the naked eye that is not uncomfortably near! Recall how many of these have also their backs distorted! Why this strange combination of deformity? Mr. Liebreich tells us: 'The greater part of it is induced while acquiring the art of writing. When the body is still being formed and is unconsolidated, the child is permitted to sit with the chest and back bent forward, and with the eyes close to the paper. Thus the natural refraction of the lenses of the eyeball is permanently prevented; the parallel rays of light are brought to a focus before they reach the retina, and there is produced the deformity of *short sight*, for the correction of which an artificial lens or glass is required. At the same time the back abnormally bent retains its abnormality, and short sight and curved spine go together, twin defects of one error which ignorance of the simplest principles permits the devoted and affectionate parent to overlook, as if it were a necessary and therefore irrepressible and irremediable evil. Let us suppose the women of our country trained to a knowledge of the first and elementary truths about visual function, and guided by them, is it not all but certain that another deformity would in a generation become virtually a physical misdemeanour of the past?

To this knowledge of nervous function it would be advisable to add to the store of elementary principles a few facts respecting the great glandular system of the body; that system which produces the digestive and other active secretions, the saliva, the bile, the pancreatic juice; which absorbs food; which takes up and, as it were, drains the tissues and eliminates those fluids and excretions by which the effete and useless animal material is removed from the body.

Of those little fleshy engines which clothe the skeleton, which are the active organs in animal motion, and which, impelled and directed by the nervous system, are the active workers, the night and day labourers of the body, the muscles, the woman should learn sufficient to be made aware of the advantages of so training the muscles to work that they shall be daily exercised, shall not be subjected to overstrain, shall be equally subjected as far as possible to healthful labour, and, by good and simple and systematic culture, form that external build of man and woman which the classic ancients of the classic age would accept as the model of the most powerful, the most symmetrical, the most beautiful of the types of the genus *homo*.

And of the bony skeleton, on which the muscular engines are laid, which forms the passive framework and supplies levers of the engine, she should gather enough information to be conversant with all its outlines of form and chemical construction. She should ascertain from her teacher that the bone, made up of two parts, an organic gelatinous part for shape and basis of support, and an earthy part for strength and durability, cannot be supplied with material for construction in unequal portions without yielding a deformed skeleton. That, deprived of its organic gelatinous part, it will become brittle and easily broken; that, deprived of its earthy part, it will be distorted, bent under the weight of the body, and yield bended limb, crooked spine, and diminutive form as the result of this one and serious deprivation of constructive material. The educated woman who has seen the exquisite build and symmetry of the skeleton; who has taken measurement of the cavities in which such vital organs as the lungs and heart are placed; who has fixed in her mind's eye the graceful curve of the spinal column; who has gathered the main facts about the sustaining parts of the skeleton; would, moreover, collect, from the physical demonstration, a series of inferences which would make her turn pale with dread and disgust whenever she detected one of her foolish sisters strangling her body in tight corset and murderous belt, to make herself hideous as well as useless, or destroying the perfect arch of her foot in a contracted foot-vice elevated on a pegtop.

Lastly, the woman should attain so much instruction in reference to the great membranous expanses as to know them also. She should study with special care that extended membranous expanse—the skin. She should learn how sensitive that surface is to heat and cold, and how, as a grand breathing surface, it gives up from its host of little sweat-glands volumes of invisible water, vapour, and gases, which, left in the body, must either be expelled by the lungs or remain to dull the sensorium and weaken the physical activity. She should learn from this the necessity of keeping the functions of the skin in due cleanliness and condition for work, so that the bath, seen to be more than a luxury, should be considered as one of the necessities of the daily life, like a daily meal of cleanly substance.

The living house thus generally learned, the Sanitarian helpmate, for us who can do so little beyond suggestion, should be tempted to study, until she completely mastered it, the mysterious construction of that deadly-lively house, which until lately the architect and builder have pitchforked into street and square with facile and contented wisdom of wigwam descent. She would require here, like Madam Ischomachus, to grasp all the details with as much precision as the old Phœnician merchant, or the modern yachtsman who knows the details of his immaculate craft so well, that even in storm, hurricane, fire or disease, all resources are ready at hand. She would require in these days to know this and something more. She would want to learn how the immaculate house is in every room provided with at least moderate ventilation. She would require to find out how most effectively and economically she can maintain in the varying seasons an even and equable temperature. She would aim to consider in what way she could keep the air of the house free of that most objectionable of mischiefs, dust. She would demand to have marked for her on a map or plan the precise position of every drain-pipe in the establishment, and would insist, with intelligent skill, on having every drain kept as systematically clean as the china in the housemaid's cupboard or the metal covers that make so many bright and effective pictures over the dresser of the well-arranged kitchen. She would see, not trusting to the mere word of any one, that those drains were properly ventilated, so that sewer air could never enter the domain except as a burglar might enter by special skill and violence, against which there is no absolute protection. She would learn enough of the chemistry of water to enable her to determine, with as much facility as she could tell whether a looking-glass is clear enough to reflect back without fault the image of her face, when a water is wholesome and drinkable; and she would have a sufficient amount of skill to direct how an impure water might be purified and made safe for her and hers to drink and use for all domestic requirements. She would see to it that sunlight found its way as freely as possible into every apartment. She would see that damp had no place in any apartment. She would insist that where any living thing that ought not to be present in a house exists in it, that house is unclean, and in some way uninhabitable for health; since health will not abide with anything that is uncleanly. She would see to the biennial purification of the dwelling, as though a Passover were still the universal belief and practice. She would make the very act of cleaning and cleansing clean; she would make the very places for cleaning and cleansing—the scullery, the landing, the bathroom, the laundry—the cynosures of the household.

In the art of perfection or towards perfection of health the educated woman would in her domestic sphere bring her best energies to understand the selection, the purification, the preparation, and the administration of foods and drinks. I have shown in two striking examples how, by a simple application of knowledge, she might prevent two great national disfigurements and disgraces of ignorance. She may go far beyond that advancement, great as it is. As she would keep seeds of certain pestilence from her fold, or vulgar poisons that kill outright, and proclaim at once with loud voice against disease, or murder, so would she do her best to keep out those refined and subtle poisons which, in and under the name of strong drinks, bring silently more accident, disease, and murder into this inscrutable world than all the other poisons put together, unlicensed though they be and so little liked by the exciseman that he would fly them any distance, the De'il himself in company, rather than so much as touch them with his divining rule.

I think, too, that in regard to foods an intelligent study based on a knowledge of the natures and uses of foods would enable her, not merely to carry out the best selections and preparations now known, but would lead her to introduce certain new and much improved methods of feeding. That she would acquire a thorough knowledge of the best art of cookery I take for granted; that she would acquire a good knowledge in choosing foods in season I take for granted; that she would become an adept in detecting actual wholesome from actual unwholesome foods I take for granted; that she would find out what foods are most suitable for persons of different age and constitution I take for granted; and that she would distribute food with well-balanced hand, neither feeding over-indulgently nor parsimoniously, that also I take for granted. But I expect she would learn to do more than all these things in relation to food, and would help, perhaps lead, in a work of the future that is in the truest sense universal in its objects. She would be able better than any one to put to the test the experience whether it is good or necessary to go to the living animal creation at all for human food. I do not wish to introduce any false sentiment into this question. It is unnecessary for me to say that every cultivated mind revolts at the sight of the shambles, and that our inner consciousness shudders when the veil is lifted which conceals the processes through which the animalised meal passes before it reaches the table. More to the point is it for me to wish to know whether it is philosophical, that is to say, truly physiological, for us to go to the inert and dead to get the best sustainment for the quick and the living. I am in doubt. It does not seem to me that man is constructed so as to be of necessity a carnivorous animal. It does not seem clear, putting the anatomical argument aside, that it can be good to go to secondary sources of supply for our food when nature bountifully presents them to us from her prime source. It does not seem reasonable that we should employ millions of living laboratories for our daily food, and take the risks of disease which they in endless forms produce and propagate for us, when we can get all that is necessary without the chance of such production and of such propagation. It does not seem certain, when we know that the vegetable world is the original source of every particle of living food, and that carnivorous animals have to depend on the herbivorous for their supplies—so that carnivorous feeding is an anomaly rather than a basic principle of nature—it does not, I repeat, knowing these things, seem certain that the cost of the support of the living laboratories is justifiable on any ground except the extravagant process of making work that work may be at hand and employment procurable. In old and barbarous times, when implements were few and animals were plentiful, it is easy to see why men should feed by hunting and by slaying; and it is easy to understand why in a becalmed sea a vigilant captain should set his restless crew to the employment of polishing an anchor. It is not so easy to see why in this day, when the great question of peace is food, cheap food, good food, healthy food, and when means for endless, refined, and ennobling employments are open, we should still maintain the practices of a barbaric era. Still I confess I am in doubt. I am not sure whether the necessity for the secondary supplies of food for man, from the animal world, are or are not necessary, and that doubt it is in the *rôle* of the educated woman to solve. Her discernment, properly and eagerly directed, would soon settle whether those about her were injured or benefited by an exclusive vegetable and fruit diet. The very timorousness which

Xenophon describes would make her study the more watchful and her experience the more definite.

However she might solve this grand enigma, sure I am that in watching carefully over food and feeding the educated woman would quickly discover a world of facts that would be of unspeakable value. It has been one of the endeavours of my life to show that we living men and women make in our own corporeal structures a refined atmosphere, which I have called a nervous atmosphere, or ether : an atmosphere which, present in due tension, distinguishes life: which absorbed or condensed distinguishes death : an atmosphere through which the external world vibrates and pierces us to the soul : an atmosphere which pure and clear brings us peace and power, and judgment and joy : an atmosphere which impure and clouded brings us unrest and weakness, and instability and misery. A physical atmosphere lying intermediate to the physical and metaphysical life : an atmosphere which our great colleague, William Crookes, might call radiant.

That atmosphere, serene or troublous, light or gross, bright or gloomy, we make in ourselves, not from ourselves, but from what we take into ourselves and transmute there. We make it from foods and drinks, and as we make it it makes us. Go into the wards of a lunatic asylum and notice amongst the most troubled there the odour of the gases and the vapours they emit by the skin and breath. That odour is from their internal atmosphere, their nervous ethereal emanation. They are mad : mad, we say, up to suicide or murder or any criminal folly. Can it be otherwise ? They have secreted the madness; they are filled with it; it exhales from them. Catch it, condense it, imbibe it, and in like manner it would madden any one! In one experiment of mine I have shown that a common product, a food if we like to call it so, a thing that can be made from food-stuff, an alcohol, will by its mere artificial temporary diffusion through the healthy body bring on, for the time it is acting as a false atmosphere, such awful despair that the experimentalist can barely avoid destroying his own life.

See, from the study of foods, out of which the radiant or deadly atmosphere is made, what fields of discovery open to the mind. A mother, watching the effect of food on her gloomy saturnine child, may detect how she can so feed it that the cloud shall pass away. Happy mother of a child! Far, far happier mother, perchance, of science and hope. In some great establishments for the insane so much gloom is secreted in the nervous recesses of human frames that many times a day, but for excessive vigilance, some terrible hand would raise itself against itself, to kill itself. What if in a wiser day, however far off, the removal of that little cloud from a troubled child should show the way to the removal of those denser, blacker clouds which lower and create storms in human breasts, overpowering altogether the radiant nervous ether! What if from that minor event this greater one should follow! What nobler accomplishment of noble deed could woman perform, save and except when she is the mother of her kind?

Referring back to our friend Ischomachus, and Madam his lady, I said he would probably not wish that she, like Hippocrates, should be learned in diagnosis. Neither in this day should I press that as a part of the education of the sanitary female scholar. I do not say this as if to frighten any one away from an art too obscure to be thought of, for diagnosis is one of the easiest and most commonplace of human acquirements when the superstitious mystery that is made to surround it is cleared away. But I say it for the reason that the art is not necessary for women except in a limited degree. I would claim, however, that to this extent it should be cultivated by women. They should know the correct names and characters of the more common diseases, and they should know, by sight, the everyday contagious or communicable diseases. To this knowledge of the communicable affections they should add a few facts bearing upon the periods of incubation of these diseases, the periods, that is to say, between the time when they are what is vulgarly called 'caught,' and the time when they are developed and in turn communicable, or, again to use a common term, 'catchable.' Thus, to know that scarlet fever may be incubated in a few hours, while small pox takes twelve days, measles twelve to fourteen days, and so on, is very useful knowledge. It enables the question of isolation of the unaffected or the removal of the sick to be rationally considered; it suggests inquiry as to the origin of the infection or contagion; and it gives reliance to those who are attending to the wants of the affected. In like manner it is well for women to know the critical periods, special dangers, and ordinary modes of termination of diseases. Beyond this, diagnostic skill, on their part, needs little further development.

At the same time all the best known methods of preventing disease should be at their fingers' ends, and the rule of the sick-room should be their faithful care. The woman should know everything about registering the temperature of the sick-room and degree of humidity; the mode of ventilation; the different special methods of feeding, washing, and changing the sick; the most efficient means of disinfecting, and of removing or destroying the poisons of the communicable diseases. How, in this way, the woman could help the physician none but the physician can understand. I have said it many times, and, on the principle that —

'Truth can never be confirmed enough,
Though doubt should ever sleep,'

I declare it again, that if, in the management and treatment of any of the acute and of many of the chronic diseases, you gave me, in this climate, absolute control of the fire and the window of the sick-room, I could determine the course of the illness. As many as you like of my learned brethren might come and go, and consult and prescribe; let me have exclusive right to those two influences, the fire and the window, and the fate of the sick man is in my hands, the best other efforts all but void and vain. How vital, therefore, the influence of the woman, educated to sanitary work, in the sick room. What an aid to the physician! Nor to the sick alone should this systematic care of the woman be directed. It should extend, more carefully than it has ever yet done, to the very young, to those who are in the first weeks and months of life; so that they be saved pains and impressions, which received and registered, if not remembered, may be penalties of after days. I conceive, in fact, there is no department of practice more neglected, in respect to principles, than the management of offspring in its earliest youth. Love there is plenty of; admiration unbounded; rational systematic training, the poorest that can be described.

I fear I am keeping you too long; let me then be content to point out but one more lesson for the modern edition of Madam, the wife of Ischomachus. She should have, in addition to instruction on all the points above named, a good training also on some subjects which refer to mental as well as physical education, and to some qualities that lie somewhat out of the way of what is purely physical, and which yet obscurely lean towards it. In these directions she should understand the little appreciated law of temperaments: the nervous, the bilious, the sanguine, and the lymphatic. She should study the combinations of these, and she should observe how temperament influences health, taste, activity, and disease. From this she would learn how different natures would intermix in work or play, and what work, what play, would suit the nature. The sanguine ruddy child, with blue eyes, red hair, strong muscle, quick movement, restless limb, she may set to study at books while she curbs exercise, with no fear that books will kill, for that child will outlive any book. The bilious child, with dark eyes, dark skin, black hair, stolid expression, thoughtful brow, she will not set to the study of books as the work of life; for books may kill, while physical exercise may save, and will never be carried, voluntarily, to injury. The nervous child, with fair skin, light hair, blue eye, quick but feeble movement, timid glance, yet perhaps unbounded ambition, she will spirit gently; will balance between physical and mental labour; will apportion excess of neither, and will never urge unduly to any effort. The lymphatic child, large of body, pale, with grey or blue eyes, brown hair, shambling step, watery lip, and slow determination, she will rouse to action both physical and mental, with the full assurance that neither effort will do anything but good.

Beyond the study of the temperaments and the special dangers connected with them, she should devote her mind to the consideration of what the learned D'Espiné has designated the mental contagions. She should study emotional contagion with special care, and on one emotion, that of fear, she should keep the most watchful observation, because she will discover it to be the most common and disastrous of all contagions. She will never excite it for a moment by story of superstition or dread. She will never suggest it. To tell a fainting or feeble person, 'You look weak, you look pale,' is, as she will learn, to add weakness to weakness, pallor to pallor, and ashes to ashes. She will lift up; disperse moral contagions wherever they are found; isolate the susceptible to them, as far as it is possible, from the centres of them; and through the windows of the mind let nothing pass but the sunshine of mirth and strength and beauty.

Finally, in physical psychological training there would stand out for contemplation, and action founded upon it, one more subject: that marvel of the marvellous in living phenomena, heredity of type and action, extending to health, and extending, alas! to disease in its deepest foundations. A little aid from books of learned men, of *the* learned man of this branch of knowledge especially—you know I can only mean Darwin—would help the scholar much; but the aid she will soon be led to find in the yet higher authority of nature will help her most. She will see the descents from good to good, and even—though fortunately with decreasing ratio—from evil to evil. She will see the conquest of death as a natural conquest over evil, and, being now in the groove of nature, she will detect how even she may availingly help nature. One effort here as a Sanitarian would call forth her strongest powers. She will stand to resist with her full persuasive might that process which I have elsewhere called the intermarriage of disease. She will tell her sisters what that terrible process means. She will tell that diseased heredity united in marriage means the continuance of the heredity as certainly as that two and two make four; that madness, consumption, cancer, scrofula, yes, and certain of the contagious diseases too, may be perpetuated from the altar; and that the first responsibilities of parents towards the offspring they expect ought to be not how to provide for wealth and position, over which they have no control, but for that preliminary wholesome parentage, which is the foundation of health, and without which position and wealth are shadowy legacies indeed. Delicate ground, you may say. I admit the fact. But in a world in which they who study the living and the dead most carefully rarely see a man or woman hereditarily free from disease, even this ground must be entered on by the enlightened scholar. I touch on it here for the best of all reasons, that the subject it includes, affecting deeply the human heart in its sympathies and affections, is one on which the influence of woman—the arbitress of the natures that are to be—is all potent for good or for evil.

To know the first principles of animal physics and life; to learn the house and its perfect management; to learn the elementary problems relating to the fatal diseases; to ordain the training of the young; to grasp the elements of the three psycho-physical problems, the human temperaments, the moral contagions with their preventions, and the heredities of disease with their prevention, these, in all respect and earnestness, I set before this Congress as the heads of the educational programme for our modern woman in her sphere of life and duty. Let these studies be hers, and once more may be applied to her the promise of that wisest of men, with whose words I opened this discourse: 'She shall rejoice in time to come. The heart of her husband doth safely trust in her.' And—sun and sum of all hopes, ambitions, happiness!—'Her children shall rise up and call her blessed.'

DRESS IN RELATION TO HEALTH.[1]

THE character of the dress of a person stands so near to the character of the person who is the wearer of it, that it is difficult to touch on one without introducing the other. All sorts of sympathies are evoked by dress. Political sympathies are in the most intimate of relationships with dress; social sympathies are indexed by it; artistic sympathies are of necessity a part of it. In a word, the dress is the outward and visible skin of the mortal who carries it.

A charming and at the same time a very useful lecture might be written on the metaphysics of dress; but in this practical day, when the useful only is tolerated and the charming is considered superfluous—I mean, of course, in a lecture—I must let all attempt at such a combination fall to the ground. I must deal only with what is purely physical: the physical body and the physical stuff that is put on it: Dress in relation to health.

In studying this subject I will consider the following topics:

---

[1] Lecture delivered at the London Institution on Monday, March 1, 1880.

Dress in relation to its mechanical adaptation to the body.

Dress in relation to season. I mean, the amount of clothing that should be worn at different periods of the year according to seasonal changes, in this English climate.

Dress in respect to the admission of atmospheric air through it or beneath it to the surface of the body.

Dress in relation to the colour of the material of which it is composed.

Dress in relation to the action of colouring substances which are introduced into its fabric and which come into contact with the surface of the body.

Cleanliness in dress.

These are all very serious subjects in respect to dress. If it were on the fashion of dress I had to treat, if I might have permission to lead you, as at a fancy-dress ball, through the historical domain of costume, then I might try to fascinate the most fastidious, and to make the time pass like a dream, in a promenade. Confined to health and dress, I can commit no ecstasy. I must be allowed to criticise, if not to scold, and rarely indeed to find one passing word that stands for commendation.

Let me, nevertheless, at once state that I have not a syllable of expression to bring forward against good fashion, and good changing fashion in dress. There is nothing whatever incompatible between good fashion and good health; they may always go well together, and they ought to go together. Naturally, I believe, they would always go together, because they are both good, and two goods can never make a bad. In like manner, bad fashion in dress and bad health go together very often, because two bads cannot make a good. For my part, I have never seen a good fashion of dress that was not a healthy fashion, and the world has only been led astray on this matter by the unfortunate circumstance that it has allowed its tastes to be directed by the childishness of ignorance. In early times costume, naturally enough, sprang out of innocence. Scientific rules were unknown, and, if we may take the history of primitive nations as true, artistic rules were not supremely developed or carried out. Through long ages fashions varied, mainly on the artistic side, approaching only towards scientific necessity in cases where arctic cold or tropical heat enforced some kind of consideration for the person who had to be clothed. Later, in more modern and scientific times, fashion has been governed by the most superficial, vain, and imprudent of so-called artistes and fashion-leaders, who have invented modes out of their own little heads, and have set Nature at defiance, as if they were Nature, and she were an idiot,—thereby changing places with her in the most complacent manner.

Let me say further even than this. I commend good fashion and fine, nay exquisite, taste in dress as a good thing of itself, independently of health. I agree entirely with Mrs. Haweis that it is the bounden duty of every woman to make herself look as handsome as ever she can. If she have natural beauty, she ought to study how to maintain it in and through every period of her life—yes—to the last; for there is nothing more beautiful than beauty in old age. If she have moderate beauty, she should do her utmost to make the best of it.

If she have no beauty, she ought to impart all that is possibly near to it, by every kind of justifiable supplement. If she be positively ugly, the more is it her duty to use every legitimate art to hide the fact, and to transform even ugliness into passable presentation. Look at an ugly woman badly attired, and showing all the lines that offend taste. Look at the same woman gracefully attired and fairly, artistically, got up, with some approach towards the beautiful, and who would hesitate to pronounce in favour of a longer *tête-à-tête* with the last of that woman as compared with the first? Why! we blockheads of men are sometimes entirely taken in by skilful ugly women. We look upon them as handsome. The deception is justifiable, and our satisfaction is more than a recompense for our stupidity.

What is good for women is not worse for men, but I am sorry to say that men are far behind women in their endeavours to assume the beautiful. In my time I have never, off the stage, seen a man dressed many removes from the hideous. When I first began to look at my male seniors, universal black was the rage, black from head to foot; the very head, which was the only part of the animal that emerged out of darkness, rising from a broad black ring called a stock, into which the chin sometimes dropped. A little later, and an extremely tight mode of dress came into fashion, a mode which is not yet entirely discarded, but which still fits closely to those strangely occupied individuals called 'copers,' about whom there is a mystery as to whether their clothes were not originally and permanently modelled to their bodies. Recently there has been some attempt at improvement in English male attire. The surtout coat, rather loosely fitting, and cut so as to hang well from the shoulders, has imported a modest but good change in fashion, while the looser and better shaped nether habiliments have so improved in design that even the sculptors have at last, with much compunction of conscience, ventured to reproduce them in marble.

Still, in the attire of men, and I think I must say in the attire of women also, a great deal is wanting in taste, and the most bigoted Darwinian would hardly, I think, dare to declare the doctrine of 'the survival of the fittest' in respect of modern clothes, whatever he might say of the wearers of them.

I name these points that I may not be accused of feeling no care for the fashion connected with dress. I would have good fashion go with every hygienic improvement in clothes and clothing, and I know it would be easy to prove that hygiene of dress could always be combined with the most artistic and perfect of fashionable designings, by which combination health, comfort, and elegance would all be insured.

Such combination set forth as a national fashion should pass, as I think, through all classes of the community, for assuredly, even at this time, though it be better than it once was, few things designate classes and keep up distinctions of classes so much as the clothes that are worn, as badges, I had almost said, of the wearers. The costumes of the trim shopman, the slovenly mechanic, the country labourer, the flourishing squire, the tight-laced soldier, the club exquisite, the lugubrious doctor, the devil-may-care artist, and the awful ecclesiastic in his demented hat and sacred pinafore—these costumes and others betray a want of national taste and national unity which I for one, health-seeker as I would be, utterly repudiate. There can be no amalgamation of mind and heart while these distinctive outside declarations exist amongst us. In robes of office, during periods of office, men may well be distinctively clad. On the bench, at the bar, in the pulpit, in the professor's chair, such costumes are classically graceful and usefully distinctive, while in the workshop or other place of business a particular outer dress suited to the occupation is no doubt necessary; but for ordinary intercourse something in common in the way of dress were surely, in these advanced days, the thing to cultivate.

I pass now to the first head of my subject: Dress in relation to its mechanical adaptation to the body.

Dress and its Mechanical Adaptation.

The first and most serious mechanical error committed on the body by dress is that of tightness, by which pressure is brought to bear upon some particular part. Presuming that an equable general pressure, not extreme in its character, and including the whole body, were applied for a fitting purpose, that is to say, for the purpose of indicating outline, no great evil probably would follow from the application of such pressure, provided that it were so adapted as to give with the growth, to yield a certain measure of elasticity, and to permit perfect freedom of motion. A little more, perhaps, may be admitted even than this. In advanced life, when the shape of the body becomes irregular, and when the weight of pendent parts drags on the rest of the body, clothing specially adapted to those parts, and surrounding them with close and even pressure, gives useful and effective support, adding greatly at the same time, it may be, to the appearance of the body. These are exceptional conditions requiring exceptional management.

That kind of pressure to which objection must be most determinately taken is where the pressure is used, not for giving support to the body, nor for sustaining natural outline, but for the express purpose of producing an entirely artificial shape and outline. It is astonishing how resolutely the advanced professors of medicine, in all times in which they have written, have denounced the practice of compressing the body in the stages of its growth for the purpose of moulding it into some unnatural form incident to fashion. It is equally astonishing to find how resolutely the votaries of the fashion have resisted the teachings of science, which may be said never to have made a single point in advance towards a practical victory; for although fashion has, now and then, given way for a short time, it seems always to have gone back into its old place.

For my part, I can do no more than earnestly follow my predecessors and compeers in their crusade against unnatural practice in dress, and especially against it as it affects the female part of the community. The corset and the waist-belt I must once more condemn as opposed to all that is healthful and all that is beautiful. By these appliances, through which an unequal pressure is exerted on one part of the body, the functions of the lungs, of the heart, and of the digestive organs are all kept under imperfect condition. The breathing is suppressed, the heart-beat is suppressed, the digestive power is suppressed. In this way the tripod of life—for life rests on the digestion, the respiration, and circulation—is made imperfect, and with that imperfection every other part of the body sympathises. Of late years women have raised the cry, and I think quite properly, that they are too much subjected to the will of men, that they have not the privileges which should belong to them as fellow human beings. But, in fact, no subjection to which they have ever submitted can be greater than this to which they have subjected themselves; and I would venture to say that, while they continue this self-infliction, they can never, under any improved system of social freedom, experience the benefit of the change. If to-morrow women were placed in all respects on an equality with men, if they were permitted to sit in Parliament, enter the jury-box, or ascend the Bench itself, they would remain under subjection to superior mental and physical force so long as they crippled their physical, vital, and mental constitutions by this one practice of cultivating, under an atrocious view of what is beautiful, a form of body which is destructive of development of body, which reduces physical power, and which thereby deadens mental capability.

Of the two evil practices to which I refer, the tight waist-belt is, I think, worse than the tight corset, except where the corset is so adapted that it acts at one and the same time as belt and compressor general. The effect of either is to press down upon the liver and stomach, to prevent the free circulation of blood through these organs, to diminish their active physiological function, to make them descend and compress the vital organs that lie beneath them, and so to impair the growth and action of all the great secreting structures. The effect, again, is to interfere with the great breathing muscle, the diaphragm or midriff, which divides the chest from the abdomen, and which, by its descent, causes the lungs to fill in breathing. Lastly, the effect is to press upwards, and so to interfere with the heart and lungs themselves. An eminent Parisian physician, M. Breschet, recorded many years ago the facts relating to a woman, who, on the right side of her throat, had a swelling which reached from the collar-bone to the level of the thyroid cartilage, and which, when the chest was tightly laced in corsets, was enlarged to its fullest. In the swelling the murmur of respiration could be heard when a stethoscope was applied over it; but when the chest was set at liberty and the swelling was gently pressed downwards, it disappeared. In this instance, a portion of the right lung had actually been forced behind the collar-bone out of the cavity of the chest altogether, into the loose tissue of the neck.

Here was a very exceptional experience, no doubt—one I have not myself seen nor found record of in this country. At the same time, I have seen very close approaches to it. I have several times known the lungs to be pushed quite out of place and compressed towards the upper part of the thorax, and I have known the heart extremely displaced by the same pressure.

That which mothers and the guardians of youth ought to know is, not only the fact of displacement of organs under pressure, not only the fact of the temporary derangement of the function of the organs, but the further and more important fact of all, as affecting the future life of the person most concerned, that under the pressure the organs implicated cannot grow so as to attain their full and complete development within the

P

period that marks the outline of growth. It is impossible, therefore, that those who are imprisoned in growth can attain full development of body. The folly they pay for in youth extends through middle age, and expedites the decline.

The evils arising from compression of the chest, as above mentioned, are not confined altogether to the female sex. They are brought about in boys and in men. It often becomes a habit in schools and colleges for youths to employ a strap or other form of belt for holding up their trousers; one boy sets the example, and the others think it right to follow; so the practice becomes general, and you find a tight line indicating pressure marked round the bodies of these youths. Fortunately, in their case, as they emerge into life, and before great mischief is done, they give up the strap and take to supporting the clothes from the shoulders, by the brace, and so they escape further injury; but, while it lasted, the injury was undoubtedly severe.

There is another and more permanent injury of this kind, however, carried out by boys, and even by men, which consists in wearing a belt for the purpose of giving what is called support. Boys who are about to run in races, or to leap, put on the belt and strap it tightly, in order, as they say, to hold in their wind or breath. Working-men who are about to lift weights or carry heavy burdens put on a belt for the same purpose, their declaration being that it gives support. Actually there is not a figment of truth in this belief. It is the expression of a fashion, and nothing more. The belt impedes respiration, compresses the abdominal muscles, compresses the muscles of the back, subjecting them to unnecessary friction, and seriously impedes motion. No boy would think of putting a belt tightly round the body of his pony if he wished it to win a race or to leap a hurdle; no working-man would put a belt tightly round the body of a horse to make it pull with greater facility a load which it was drawing. On themselves they commence the practice because somebody has set the example, then they get accustomed to the impediment, and think they cannot get on without it. Drinking is learned by just the same absurd process.

I had a good working-man in my employ who would undertake no vigorous effort until he had tightened his belt. Once I got him to test what he could lift with and without the belt, and he was obliged himself to admit that he could do more without it than with it; but, he argued, he could not get on without it. That is what ladies say about corsets.

Respecting this belt for boys and men there is a word more I must say which is of serious import. When they put on the belt for the sake of performing some feat of strength, they effect another dangerous mischief. Compressing the abdomen, they force, during the exertion, the contents of the abdominal cavity downwards under pressure, giving no chance to resilience back again after the exertion or shock. In this way they frequently cause hernia or rupture. I have seen, professionally, several instances of this occurrence in boys, and amongst workmen who wear belts this accidental disease is so common that it is the rule rather than the exception to find it present.

Other forms of tight pressure upon the body are open to serious, if not to equal objection. The wearing of shoes which compress and distort the feet is a singularly injurious custom. Suppose I said that nine-tenths of the feet of the members of an English community were rendered misshapen by the boots and shoes worn, the statement would seem extreme, but it would be within the truth. The pointed shoe or boot is the most signal instance of a mischievous instrument designed for the torture of feet. In this shoe the great toe is forced out of its natural line towards the other toes, giving a reverse curve from what is natural to the terminal part of the inner side of the foot, while all the other toes are compressed together towards the great toe, the whole producing a wedge-like form of foot which is altogether apart from the natural. Such a foot has lost its expanse of tread; such a foot has lost its elastic resistance; such a foot has lost the strength of its arch to a very considerable degree; such a foot, by the irregular and unusual pressure on certain points of its surface, has become hard at those points, and is easily affected with corns and bunions. Lastly, such a foot becomes badly nourished, and the pressure exerted upon it interferes with its circulation and nutrition. It ceases to be an instrument upon which the body can sustain itself with grace and with easiness of movement, even in early life; while in mature life, and in old age, it becomes a foot which is absolutely unsafe, and which causes much of that irregular, hobbling tread which often renders so peculiar the gait of persons who have passed their meridian.

It sometimes happens for a time that these mistakes in regard to the boot and shoe are increased by the plan of raising the heel and letting it rest on a raised impediment of a pointed shape. Anything more barbarous can scarcely be conceived. By this means, the body, which should naturally be balanced on a most beautiful arch, is placed on an inclined plane, and is only prevented from falling forwards by the action of the muscles which counterbalance the mechanical error. But all this is at the expense of lost muscular effort along the whole line of the muscular track, from the heels actually to the back of the head; a loss of force which is absolutely useless, and, as I have known in several cases, exhausting and painful. In addition to these evils arising from the pointed and heeled boot, there are yet two more. In the first place, the elastic spring of the arch being broken by the heel, the vibration produced by its contact with the earth, at every step, causes a concussion which extends along the whole of the spinal column, and is sometimes very acutely felt. In the second place, the expanse of the foot being limited, the seizure of the earth by the foot is incomplete both in standing and in walking, so that it becomes a new art to learn how to stand erect or to walk with safety.

Another form of constriction in dress is that produced by the garter. By this pressure a line of depression is often produced quite round the limb below the knee, and the course of blood through the veins from the foot and leg, into the body, is seriously impeded. This is one cause of varicose veins, sometimes an original cause, and always a serious impediment to recovery when, from any other reason, the enlarged or varicose vein is already present. The ligature or band called the garter is bad in any way, but is far worse when it is worn below than above the knee, for above the knee the two tendons, commonly called ham-strings, receive the pressure of a great portion of the bandage, and act as bridges to the veins which pass beneath.

In men I have seen mischief from the tight cravat and collar, the pressure caused by the same leading to an obstruction to the due return of the blood from the brain. This, in persons of plethoric habits especially, is a danger not to be disregarded, and, though it may be of comparatively rare occurrence, it is worth mentioning. I have more than once in my life had occasion to see the injurious results produced by it.

I have now referred to the four varieties of pressure which are the most injurious in dress : pressure at the waist ; pressure at the foot ; pressure round the leg ; and pressure round the neck. I place them in the order of their importance, but the first undoubtedly outweighs the others altogether.

It is actually impossible to overstate the physical injuries which result from these mistakes in bodily attire. I have told some of them. I reserve one which I will state before I pass to a new section. It will perhaps influence some who are comparatively thoughtless on this subject; it will, I am sure, influence all sensible and thoughtful people. It is this observation, that the mischiefs inflicted by mode of dress become hereditary in character. I do not mean to say that because a person produces in himself or herself a deformed waist, or foot, by dress, that therefore that particular deformity will be physically hereditary in the offspring of such person. I think the evidence is rather against that view, because it would seem that the Chinese children, born of mothers whose feet have been mechanically distorted, are born with feet which would come to a natural condition if they were not bandaged in infancy in the same manner as the mothers' were. But of this I am sure, that the tendency to commit these deforming acts is often hereditarily received and hereditarily transmitted, and that the sense of desire for the performance of the act is also transmissible. This, in fact, is one of the great difficulties which we teachers have to overcome. We have to fight against inbred proclivities, which are so deep rooted that if all the women of England at this time could, by a voluntary act of education, be led to give up tight-lacing, another generation, perhaps two generations, would have to live before the practice was entirely abolished.

The lesson we have to learn and practise in respect to the mechanical arrangements of dress so far is that every plan which leads to irregular tightening of the body should be given up. The corset and waist strap should especially be abandoned, and our young girls should be taught to grow up just as their brothers grow, without ever learning the sense of false support which the corset soon suggests as a necessity. With the members of both sexes a reform should be introduced in the matter of boots and shoes. The tight boot should be entirely discarded, and that boot preferred which approaches nearest in form to the natural foot. Mrs. Haweis and others have insisted on the removal of the raised heel altogether from the boot, with which I entirely agree. Anatomically and physiologically it is a complete mistake to have the heel raised from the ground beyond the level of the palm of the foot. The moment the heel is raised, the plan of the arch is deranged, and the elastic wave-like motion of the foot impeded. The arch ought always to have full play, and Mr. Dowie's plan of introducing an elastic connection or band across the arch, so as to allow it freedom, is an admirable device.

The method by which clothes should be supported on the body is another extremely important subject in connection with dress, and especially in relation to the dresses worn by women. Copying probably from an Eastern custom, and from the primitive method of wearing a girdle, it has become a habit endorsed by long centuries of use for women to carry all their long flowing robes from the waist. These, tied one over the other, layer upon layer, and with sufficient tightness to enable the garments to be borne by the actual pressure upon the waist, are as great an encumbrance to the wearer as the corset. Indeed, it is sometimes argued that the corset is necessary in order that the pressure may be sustained, the corset itself acting as a kind of shield between the body and the bands, and acting also in some way like a shoulder for supporting the bands. When the dresses which are thus sustained are short and of light texture, the weight and encumbrance are considerable; but when the dresses are long, when they trail on the ground, and when they are made of heavy material, the weight and encumbrance are drags on the life, which I suspect the strongest man could not sustain while engaged in his ordinary avocations.

I am rejoiced to see that ladies themselves, who are writing intelligently on this topic, are earnestly teaching in respect to it what is both common sense and common humanity. I agree with these that the tax of carrying clothes from the waist is utterly unjustifiable, and that the parts that should bear the burden are the shoulders, and none other. In this regard women ought to be placed under just the same favourable conditions for movement of the body as men, and the greatest emancipation that woman will ever have achieved will have arrived when she has discovered and carried out this practical improvement.

In saying this I do not for a moment wish to suggest that the outward appearance of the feminine dress should be like that of the masculine dress. To the woman, the flowing robe which even trespasses a little on the ground is most graceful, and is signally characteristic of feminine beauty. I would, therefore, that it should remain in all its gracefulness; but in so far as everything else is concerned, for every circumstance in which health is involved, for warmth, for freedom of movement, for mode by which the dress is carried from the shoulders, I would say, Let the women have all the advantages which now belong to men.

For any one who will for a moment think candidly must admit that the dress of men, however bad it may be in taste, or in whatever bad taste it may have been conceived, is, in respect to health, infinitely superior to that of women. In the dress of the man every part of the body is equally covered. The middle of the body is not enveloped in a number of close layers, while the lower limbs are left without close clothing altogether. The centre of the body is not strained with a weight which almost drags down the lower limbs and back. The chest is not exposed to every wind that blows, and the feet are not bewildered with heavy garments which they have to kick forward or drag from behind with every advancing step. The body is clothed equally. The clothing is borne by the shoulders; it gives free motion to breathing; it gives freedom of motion to the circulation; it makes no undue pressure on the digestive organs; it leaves the limbs free; it is easily put on and off; and it allows of ready change in vicissitudes of weather. These are the advantages of modern attire for the man, and all I claim is that they should, by faithful copy, be extended to the woman, with the one exception of the graceful outer gown or robe, as a supplement to her own superior grace and beauty.

It is reported of the eminent surgeon Mr. Cline, the teacher of Sir Astley Cooper, that when he was consulted by a lady on the question how she should prevent a girl from growing up misshapen, he replied, 'Let her have no stays and let her run about like the boys.' I gladly re-echo this wise advice of the great surgeon; and I would venture to add to it another suggestion. I would say to the mothers of England, Let your girls dress just like your boys, make no difference whatever in respect to them—give them knickerbockers, if you like—with these exceptions: that the under garments be of a little lighter material, and that they be supplemented by an outer gown or robe which shall take the place of the outer coat of the boys, and shall make them look distinctively what they are—girls clothed *cap-à-pie*, and *well* clothed from head to foot.

In speaking of these mechanical arrangements of dress I have as yet made no mention of the throat and the head as parts requiring to be clothed. In suggesting that girls should be clothed as fully as boys I have incidentally conveyed that the chest of the girl should be covered, and I would add that in both sexes the throat should be covered also during the period extending from October to April. The throat is one of the most important parts to protect, and it is, as is well known, one of the most common parts of the body to become affected during cold weather. In this past bad weather it has been my constant—I had nearly said, daily—observation to see some affection of the throat, attended with cold, and so often has this occurred in those whose throats have been uncovered as compared with those who have used careful moderate covering, I cannot doubt that the absence of such covering has had a very deleterious effect.

Of coverings for the head, I should say that they should be always light and free, whether a bonnet, or a cap, or a hat be the subject under dispute. I think the gipsy hat beats the quaker bonnet for the fairer sex; and although for men I cannot say anything in favour of the tall chimney-pot that will redeem it from its ugliness, I must claim for it that, when it is light and well ventilated, it is healthy. The felt hats are too closely-fitting, though some are becoming. The stiff felt hat, with narrow turned-up brim, and which looks like a Roundhead's helmet without the metal, is in respect to health miserable, and in respect to appearance simply hideous. The most graceful of all head-dresses for either sex—and it suits either—is the fine old Geneva cap, sometimes called the 'Leonardo da Vinci,' which I wear on occasions, by right, as the doctor's cap of the old University of St. Andrews. It is not merely a handsome head-dress, it is healthy also, and adapts itself to heat and cold. I, for one, would willingly give up the particular privilege of wearing it to see it more widely adopted.

Dress in Relation to Season.

From the subject of mechanical adaptations of dress I pass to the amount and kind of clothing that should be worn at different periods of the year.

On this subject there is great contrariety of opinion, and perhaps still greater contrariety of practice. There are those who maintain that to be healthy the body should be hardened by exposure to cold, and that to wrap up and coddle is the weakest and worst of all plans. It must be admitted that there are some persons who seem to flourish under this *régime*, and who live to advanced age without suffering from cold even when lightly clad. I have known myself three men who have approached their ninetieth year, and who always vigorously refused to wrap up at all. Such persons are great examples, but they are too exceptional to be counted as safe ones. The majority of the aged die, as a rule, rapidly during cold weather. I have known children who have lived through their childhood half clothed in coldest seasons; and these are great examples, but they also are too exceptional to be accepted as safe examples. As a rule, ill-clad children suffer intensely in cold weather, and often die, from cold as the indirect cause.

On the other hand, no doubt, some persons do greatly over-encumber themselves with clothes; and it is curious to observe that stout persons, who are wrapped and thoroughly lapped in their own subcutaneous non-conducting layer of fat, and who are generally feeble, encumber themselves with more clothes than their lithe and spare-ribbed friends, who really require most protection.

The truth is, that extremes on both sides are bad, and that a dash of good common-sense is required to equalise them.

In this climate the regulation of dress in relation to health is an actual necessity during the varied seasons that prevail. We may take it as a general rule that when the body requires more food and more sleep to meet the cold, it requires also more clothes than it does at times when sleep and food are also less wanted. There is a very remarkable physiological truth bearing on this point which every one ought to know, inasmuch as a knowledge of it becomes a guide to us in our daily life, not only in relation to dress, but to food, exercise, labour, and repose. The truth is so practical, that I dwell upon it with some detail. It is this. There are certain periods of the year, in this climate, during which, independently of our wills or our actions, we are gaining in bodily weight, while there are other periods when we are losing, both periods showing a regularity which is as singularly correct as it is singularly interesting. This truth was first discovered by my late friend, Mr. W. R. Milner, for many years medical superintendent of the large prison at Wakefield. His discovery was elicited by the laborious process of weighing, daily, immense numbers of prisoners through various seasons for a long series of years. I give his results as he himself has stated them.

The prisoners were all males between the ages of sixteen and fifty, and were presumed to be in good health when committed. The cells in which they were confined had a cubic capacity of about nine hundred feet, and from thirty to thirty-five cubic feet of air were passed through each cell per minute. The mean temperature of the cells for the entire year was 61°; the highest monthly mean, 66·5°, occurred in August; the lowest, 56·9°, in March.

The diet was uniform, with the exception of the alterations ordered by the medical officer in individual cases, and consisted of the following articles daily: Bread, twenty ounces; meat without bone, four ounces; soup, half a pint—these are equivalent to about seven ounces and three-quarters of butcher's meat; potatoes, one pound; skimmed milk, three-quarters of a pint; gruel, one pint, containing two ounces of oatmeal.

The dress was a cloth jacket, waistcoat, and trousers; cap and stock; linen shirt; woollen stockings; drawers, and under-shirt.

The prisoners were sent out to exercise in the open air nine hours a week; the exercise was for one hour at a time; the men walked in circles, and every ten minutes they ran for a hundred and fifty yards. They were all supplied with work, and were for the most part employed in making mats and matting of cocoa-fibre and other materials; some worked at tailoring and shoemaking, and a few had other work to perform.

All the prisoners were weighed on admission, and at the latter end of every calendar month during their stay.

The number of prisoners over whom these observations extended was four thousand; the period of time occupied, ten years; the average number weighed monthly, three hundred and seventy-two; and the total number of weighings, forty-four thousand and four.

The men had all been weighed by Mr. Milner or under his superintendence, and the series of observations were unbroken.

The results of these weighings were tabulated on various bases, with a view to isolate the effect of a certain number of variable bases on the gain or loss of weight among these prisoners, and to determine the amount of influence exerted by each of these conditions.

The conditions selected for investigation were :—
1. The season of the year.
2. The period of imprisonment.
3. The employment in prison.
4. The age of the prisoners on admission.
5. The height of prisoners on admission.

The influence exerted by each of these conditions was well marked, and, with one exception, viz. the influence of season, the deductions were such as would have been anticipated.

The first showed the influence of the season of the year on the weight of a number of men placed during the entire year under circumstances of food, clothing, and work which did not differ, and who, for the greater part of the day, were in a temperature which did not vary greatly between the hottest and the coldest months. Under such circumstances it might be expected that the weight of the men, taken as a whole, would remain sensibly the same; and that the numbers losing or gaining, as well as the quantities lost or gained, would vary little month by month; or that, if any marked variation occurred, it would be of an accidental character, depending on the greater or lesser amount of sickness during any particular month. The results, however, showed that a marked periodicity existed, and that, taking an average of years, there were two distinct series of months, during the one of which there was a constant loss of weight, and during the other a constant gain, so that, if the year were divided into quarters, there was a loss during the first and fourth quarters, and a gain during the second and third.

The two series of gaining and losing months were unbroken, except in one instance. On reference to the results it was found that in November, which was in the losing series, a gain occurred. The amount gained was very small, and the discrepancy was caused by the arrival of large numbers of prisoners in September and October, who usually gained weight for a short time after they were received, so that probably this break in the series resulted from the influence of the stage of imprisonment, which rather more than balanced the influence of season. On estimating carefully the facts which showed the average gain or loss per prisoner weighed, it was seen that, beginning at December, the amount lost per man increased rapidly, and very steadily till March, but that between March and April there was a very abrupt transition from loss to gain. The gains then continued till August, the amount gained increasing on the whole by a series of jerks, each alternate month presenting a larger and a smaller gain respectively; so that, to obtain a steadily increasing series, it was necessary to couple the summer months in pairs. Between August and September a change of weight occurred, about equal in amount, but in the opposite direction, to that which took place between March and April. The changes between March-April and August-September were far greater in amount than the changes which took place between any other pairs of consecutive months; and this remark applied with greater force to the percentages of men gaining or losing, and to the net gains and losses per man.

The inferences which may be fairly drawn from these observations are:—1. The body becomes heavier during the summer months, and the gain varies in an increasing ratio. 2. The body becomes lighter during the winter months, and the loss varies in an increasing ratio. 3. The changes from gain to loss, and the reverse, are abrupt, and take place about the end of March and the beginning of September.

The results, which were thus gathered from the study of a large number of periodical weighings, presented a remarkable relation to the facts obtained by Dr. Edward Smith from a series of most valuable and elaborate experiments which he made on the quantities of carbonic dioxide thrown off by the lungs at various seasons of the year. For instance, Dr. Smith found that the quantity of carbonic dioxide thrown off was much greater in winter than in summer. Milner's weighings showed that the prisoners lost weight in winter, when the evolution of carbonic dioxide was great, and gained weight in summer, when less of that gas was exhaled.

This in itself would be a striking coincidence; but it was clearly detected that a sudden change took place between March and April, and at the same time of the year Dr. Smith found that a similar change took place in the amount of carbonic dioxide thrown off, and that the *amount* of the change was much greater at that period than at any other time; and so much greater, that the alteration struck him as being a very remarkable circumstance. Dr. Smith's observations corroborate those of Mr. Milner by shewing that a more active combustion takes place in the body with consequent increased waste in winter than in summer. There can be little doubt that variations of temperature, and of light, are the principal agents in causing these changes; but it will probably be found that, in addition to the direct influence of these physical agencies, a periodic action in the system adds to or diminishes the effect.

From the consideration of the facts collected we may fairly infer that there is a periodic variation in the weight of man during the year, the six summer months being gaining and the six winter months being losing months. The amounts gained or lost gradually increase from the commencement till the termination of each period respectively; the change from the gaining to the losing period, and the converse, is, however, abrupt, and these changes take place at times not very distant from the vernal and autumnal equinoxes.

Bearing on the question thus raised by Mr. Milner, I myself, from the Registrar-General's returns, made an analysis of 139,318 deaths occurring, from 1838 to 1853, in London, Devonshire, and Cornwall, with a view of determining what causes of death were connected with the varying seasons of the year; and the result was to discover that during the wasting season, which was by far the most fatal, those diseases were most rife which spring from exposure to cold, and which are extremely fatal under that condition. I have since then many times drawn special attention to the importance of regulating clothing so as to meet the emergency to which the body is exposed during the wasting period; and the rules I had then in my mind I would enforce now. It should be a settled practice with all persons in these islands that they commence to put on warmer clothing a little before the wasting period begins, and that they continue it considerably beyond the time when the balance turns, and the period of increasing weight commences.

Bearing on this important point, I have received a most practical note from the Rev. B. A. Irving, M.A., head master of the College, Windermere, in which the argument set forth above is fully confirmed. Mr. Irving indicates, from meteorological data, that about the 10th of May and about the 10th of November there is a remarkable fall in the mean temperature. The fall, commencing in November, continues to increase until the end of February. The pinch of cold in May is followed by warmth, which continues through the summer. The rule Mr. Irving deduces from these physical facts is that we should be warmly clothed from the end of January to the end of February, and that summer clothing should on no account be assumed until the cold pinch about the 10th of May is well passed—say about the 15th of May. The summer dress may then be continued until the end of September ; but winter clothing should be most carefully assumed before the cold pinch of November 10th—say by the 1st of November. With this sound advice I entirely agree.

Need I hesitate to say how dangerously these simple rules are ignored, and that, too, by those to whom they most solemnly apply! The delicate girl invited to the ball or evening party, in the winter season, goes there with a throat and chest exposed or partly covered, and with all her garments as light as fashion will permit them. She goes into a close room, heated to 65°, or it may be 70°. She dances herself into a glow, and then, exhausted, excited, and breathless, she passes out of the room, to exchange its warmth for a temperature of 35°, or lower—perhaps below freezing point. She takes cold, she suffers from congestion of the lungs, and, if her tendencies are in that direction, she passes into consumption. And who shall wonder?

As spring advances, dangers increase to everybody. The weather is treacherous; a bright day or two in March seems to herald summer, and the warm clothing is cast aside. Suddenly, there is a fall of temperature with a bitter east wind, and the unprepared are caught as if in a trap. They have passed the long wintry ordeal before which so many have succumbed, and they are reviving, but have not revived. In this condition they are stricken with disease, often fatally. If we study the Registrar-General's returns through the month of March, April, and the early part of May for a few years, we see how solemnly correct is the history I am now bringing under notice.

You will ask, What kind of clothing is best to meet the varying changes? I answer, That which combines lightness with warmth, and which absorbs the watery secretion from the body without retaining it. For under-clothing I give a decided preference to silk, basing this preference entirely on practical grounds. Knitted or woven silk is at once the material which best maintains warmth, affords lightness, and transmits perspiration. If the expense of it be urged on one side, its extraordinary durability may be named as a set-off. The silk should be worn next to the skin. Over the silk, for nine months in the year at least, there should be a woollen covering which should include the whole body.

This should not be made of thick, heavy flannel, for thickness and weight contribute little to warmth, but of soft, light, fleecy material, or of that thin flannel which somewhat resembles silk in structure. The feet coverings within the shoes should be of the same character, and long socks should be preferred to stockings. The upper clothing, like the under, should be of light and, at the same time, warm character, and the overcoat or cloak should vary with the season. In coldest weather a fur coat fitting loosely to the body is the best.

Ventilation of Clothing.

Connected with this part of my discourse, there comes in naturally the ventilation of clothes on the body, to which I referred in the opening paragraphs. I cannot too seriously express the necessity of maintaining a free ventilation. Whatever impedes the evaporation of water from the body leads, of necessity, to some derangement of the body, if not to disease; for the retained moisture, saturating the garments, produces chilliness of surface, and checks the action of the skin. Then follow cold, dyspepsia, and, in those who are disposed to it, rheumatism. For these reasons I always hold that the so-called waterproofs are sources of great danger, unless they are used with much discrimination. It is true they keep the body dry in wet weather, but they wet it through from its own rain; and when the body is freely exercised and perspires copiously during rain, shut up with its own watery secretion on one side of the waterproof covering, and chilled by the water that falls on the other, it is in a poor plight indeed. It had better be wet to the skin in a porous clothing. Hence, I would advise that the waterproof should only be used when the body is at rest, as when standing or sitting in the rain. During active exercise a large, strong umbrella, not a finikin parasol pretence—is worth any number of waterproofs.

Colour of Dress.

The colour of the dress is another practical point of considerable moment. The *Lancet*, a few years ago, was very much criticised for suggesting that in the cold, dark weather dresses of light colour should be worn. The *Lancet*, nevertheless, was right. The light-coloured dress is at once the warmest and the healthiest. In the arctic regions white is the prevailing colour of the animal that most requires warmth. The same colour is also best adapted for summer wear, for that which is negative to cold does not absorb heat. The objection made to white clothing is that it soon becomes dirty, or, correctly speaking, that it shows the presence of dirt more quickly than darker fabrics. This might be an advantage in many cases, but I think it is fair to admit that white out and out, for all times and seasons, is not practical. The best compromise is grey, and I wonder that in our climate that practical fact, which was once known and acted upon, has ever been allowed to die out. Those wise and discerning forefathers of ours, who utilised the serviceable grey suits, were best informed after all in the matter of colour of dress, for health as well as for service.

Fashion, in these later times, has misled once more, by the introduction of the incorrigible black clothing for the outer suits of men and women. The inconvenience of this selection reaches its height in the infliction it imposes on those poor ladies who, after bereavement, think it necessary to clothe themselves in unwholesome folds of inky crape. Next to the Suttee, this seems to me the most painful of miseries inflicted on the miserable. Happily, it is, I think, in its last days.

Colouring Substances or Dyes.

I would make, in one or two sentences, an observation on the colouring substances that are sometimes introduced into dress, and their relation to health. When the aniline colour stuffs were brought in for dyeing under-garments of red or yellow colour, the dyes caused, sometimes, where they came into contact with the skin, a local irritation, and now and then even some constitutional derangement. The agents which were at work to produce these conditions were the poisonous dyes called red or yellow coralline. The local action of both these poisons is sharp, and they bring upon the skin a raised eruption of minute round pimples, which I have known to be mistaken for the eruption of measles by the unskilled in diagnosis. The irritation which attends the rash is painful, and if there be much rubbing of the part little vesicles may form and give out a watery discharge. Once I knew an eruption on the chest, attended with much nervous prostration, caused by a red woollen comforter; but, as a rule, the evil is purely local, the colouring matter being not readily absorbed by the skin. This is fortunate, for the poison would be intense if it were to enter the blood.

It is necessary at once to remove all coloured garments when they are causing local mischief, and such garments should never be worn until they have been many times rinsed in boiling water. As a general rule coloured clothing should never be worn next to the skin.

Cleanliness in Dress.

Cleanliness in dress, the last passage in my programme, is one on which, to an educated audience, I need not dwell. Health will not be clad in dirty raiment, and those who think it will very soon find themselves subjected to various minor ailments—oppression, dulness, headache, nausea—which in themselves and singly seem of little moment, but which affect materially that standard of perfect health by which life is blithely and usefully manifested. The want now most felt amongst the educated, in our large centres, is the means for getting a due supply of well-washed clean clothes. The laundry is still up a tree, and when you climb to it, it is rarely found worth the labour of the ascent. In London, at this moment, a thousand public laundries are wanted, before that cleanliness which is next to godliness can ever be recognised by those apostles of health who feel that their mission in the world stands on the first list of goodly and godly labours for mankind.

THE POVERTY OF WEALTH.

Two events in my life have suggested the title and subject of this essay.

In 1855, in a printing-office in Great Queen Street, Lincoln's Inn Fields—No. 37, to be very precise—the cover, in proof, of the first English journal on sanitation lay before me for revision. In the lower part of the page a picture of Hygeia, the goddess of health, was drawn; and the picture, looking very weird and lonely by itself, wanted a motto. I invented several, but not one to my taste. At last there flashed across my mind the original and happy one: *National health is national wealth.* I fixed on this at once, and in a few days it came forth as the motto of the 'Journal of Public Health and Sanitary Review.'

The motto from the first took well with the public, and it was no slight gratification to see it pass, as it soon did, into one of our best-known and most frequently repeated proverbs. I have been led since, nevertheless, to look upon it as an inverted proverb, and to give it an entirely different meaning from that which was passing in my mind when the idea, expressed originally in the five short words, presented itself. How this change came about may best be described in the narrative of another incident.

One day, long after the period named above, I was entering for professional duty the house, or mansion, of a very wealthy person. It was on a Sunday afternoon of a damp and cheerless London day. On the steps leading to the house there sat a man in the lowest possible stage and state of destitution. He craved of me that I would give him a trifle to enable him to break his fast. He had walked, he said, from Northampton on one meal and no bed. He entered into his many grievances, without any reference to misfortunes or to opportunities. His mind was a scene of complaint against home, country, friends, himself, life. He wanted food; he wanted drink still more urgently; but he did not pity himself nor bemoan his fate. He had come to a point of poverty when he did not care what happened to him. He could not be worse whatever might occur; if the world itself came to an end next minute, it would not '*sinnify*' to him; it would be rather just the thing most likely to wake him up and give him something to look at that was worth seeing. As that grand event was not likely to occur, the next best luck was a copper or two and direction to the nearest workhouse. In both the last particulars I gratified him, and bidding him good-day, saw him slink off without a word of thanks, and without exciting any feeling on my part that the thanks were either desirable or needful.

I next entered the house at the door of which I had stayed. It was one of those dark, richly furnished, warm, silent, snug, and tasteless sepulchral mansions in which the envied rich so often loves to dwell. The staircases had their steps so thickly carpeted, not a footstep on them could be heard, and through the whole place there was no sound save that of the timepieces on each landing, which ticked away in melancholy measured vibrations, as if they were everlastingly saying, one after another, 'Keep quiet,' 'Keep quiet.' 'Great wealth,' 'Great wealth.' 'Don't laugh,' 'Don't laugh.'

I entered a big saloon chamber, with that last ticking in my ear, to discover, at the far end, sitting on the window seat, another man, so entirely like the man I had met on the doorstep, that if there had been time for the transformation I should have felt sure that the man had got into the house before me, made a slight change of raiment by putting on a rich dressing robe and a pair of furred slippers, and had reappeared. The expression was the same; the dreary sound of the voice the same; the first exclamation, 'What can you do for me?' without a previous word of ceremony or greeting, was all but the same. I was literally startled. I stood before a man so wealthy that the golden calf itself might have called him brother, and I found a repetition of what I had left on the step of his door. Strangely, too, to some, but not to me, I listened to the same story of grievances, to the same views about life and its utter worthlessness; the same absolute recklessness in respect to the future; the same desire, if such it may be called, for some impossible gigantic event to bring a moment's wonder; the same dull, thankless expression for the receipt of an assistance equivalent, in its way, no more, no less, to two poor coppers and a direction to the workhouse.

Then I left, having discovered, as honest John Bunyan says, that there is a bye-way to hell even from the gates of heaven.

Henceforward, from that time I have felt that my original proverb was inverted. In that saying I treated wealth as if it was the source of health. The mistake was basic: the proverb should be, *National wealth is national health*, and there is no other.

From that time forward, too, I began to conceive the idea, from nature, of the poverty of wealth, which is my text of to-day. I descried that wealth, in the full sordid sense of that term, must lead to poverty, as day leads to night; and that if mankind, struggling at the present in the throes of socialistic revolution, would learn the right way to avoid catastrophe of revolution, it must learn the natural law which reduces extremes of poverty and wealth to synonymous terms and equal qualities, and shun both with equal determination.

The social student will say at once that this is no new conception. It does not pretend to be: there is nothing new under the sun, and never can be. Because all that is under the sun is of it, and even the sun itself may not be new.

More than that, it is all epitomised in the Book of Wisdom:

'There is a sore evil which I have seen under the sun, namely riches kept for the owners thereof, to their hurt.

'But those riches perish by evil travail, and he begetteth a son, and there is nothing in his hand.

'As he came forth of his mother's womb, naked shall he return, to go as he came, and shall take nothing of his labour, which he may carry away in his hand.

'And this also is a sore evil, that in all points as he came, so shall he go, and what profit hath he that has laboured for the winds?

'All his days also he eateth in darkness, and he hath much sorrow and wrath in his sickness.'

These reasonings, thus familiar, support generally the truth of the paradox set forth in the words, the poverty of wealth. I wish now to treat of this truth, not as a new subject, but from a different side of an old subject, on what may be called a physician's side, by which it is argued that health of body and of mind is the only true standard of wealth, and that as there can be no health in the lap of poverty or in the lap of luxury, so wealth is poverty and leads, as in the two cases cited above, to the same condition.

Let me, before proceeding further, be explicit as to terms. By wealth I mean great possession, beyond all the possible wants of the possessor, and so much beyond his wants that if he wishes to apply that which he holds as surplus over his own most liberal requirements, he must make a study and labour of the distribution in order to be so much as moderately sure that he is doing good. I do not, therefore, include under the term *wealth* what may be called a fair competency, or such possession as shall place a man and his family above the risk of want. On the contrary, I rather hold with the witty Rochester, that every man should save a competency in order that he may remain honest. Again, by poverty I do not mean frugality, or any simple and inexpensive mode of living, by which a man can live by his own exertions or savings. I cease to include poverty until it becomes so extreme that the assistance of others than the man himself is called for in order that he may be able to live.

In a word, I define a wealthy man as one above the

line of highest competency, and a poor man, or rather a poverty-stricken man, as one below the line of lowest competency. None within those lines concern us now.

With these definitions, I proceed to show, as I feel from natural observation of natural fact, that wealth and poverty are, in effect, the same; I mean in respect to the fruits they bear, their practical results on the every-day life, and their influence on the history of a nation such as ours.

I. By the common voice of mankind the word poor is applied under similar circumstances to poverty and wealth alike: we say a *poor creature* when we are speaking of any one who is out of means or estate. In like manner of a man, however rich he may be, we also speak if he be low in mind or body. We never speak of a rich creature. That man on the doorstep of the house I referred to, and that man in the mansion, were alike poor to the common observation in this sense: they were both *poor creatures*. Their poverty was, moreover, largely of the same kind: poverty of mind and body. So the term is quite correct.

II. Poverty causes poverty. 'If thou art poor, thou wilt always be poor; riches are only given to the rich,' is one of the old Latin proverbs. The poor man drags down the poor man. To him it becomes certain that, as he cannot live on his own industry, he must live on that of others. From this natural fact springs the further fact, so often commented on, that charity to the poor, misapplied, is an evil rather than a good. From this also springs the equal fact that not to help the helpless poor is heartless. In these matters the reason and sympathy of our common nature come into conflict. Reason says, 'The charity is thrown away: the help had better be bestowed on worthier objects.' Sympathy says, 'Not help the poor creature? Why, what then is he to do but steal or starve, for see, he has no capacity for labour?' Thus the brain contends with the heart, and the necessary end of the contest varies according as the person in whom it occurs is most potential in brain or in heart; and as sympathy forms the foundation of all initiative action, poverty by its claims on all above it, and most upon those nearest it, leads to poverty and feeds it. In like manner, wealth makes poverty. The physiological influences at work are precisely the same. Why should poverty labour if it can be sustained without labour? Why should wealth labour when it gets all it wants, and more than it wants, without labour? So springs forth the helplessness of wealth. Why go to the absurd trouble of spending when you have all you want without the trouble? Let the treasure take its own course and accumulate; in that there is the silent pleasure of the accumulation without exertion. Thus wealth, which should cease to be wealth, by being distributed largely for general competency to the many, becomes poverty to all whom it does not assist to competency, and to none so much as to those who hold it, and in turn are held by it in hopeless bondage.

III. It is one of the outcomes of poverty to lead its victim to low pursuits, and to wean him from all noble and great objects. There is a poverty so extreme that into the mind of its holder nothing whatever can be instilled that is intellectual or progressive. The brain is too apathetic to receive; the senses are too restless and indolent to focus or concentrate. Whatever, therefore, most pleases, must be something easy and light, such as an animal might take delight in. The lower organisations proclaim themselves dominant. To eat! yes, ravenously; to drink! yes, to any degree; to smoke! yes; to look listlessly over some remarkable sight! yes; to hear or read some story that costs no tension of mind! yes; to repeat the same stories over and over! yes; to find the warmest and cosiest place in which to sleep, and to sleep on as long as possible, without disturbance! yes. All this is the luxury of life. Poverty of poverties!

But wealth leads to the same poverty: to eat, to drink, to smoke, to revel in inane pleasures; to turn listlessly from one thing to another; to repeat the same old stories and fooleries; to speak in an idle drawl; to sleep in the warmest and cosiest quarters; to indulge in sleep for the longest of possible times, these are the poverties of wealth, in the full exercise of the faculties of poverty.

IV. We are constantly reminded that poverty presses men into dishonesty. It is the fact, and in certain examples of the extremest kind it presses with such pressure that the natural sentiment expressed in the words 'Necessity knows no law' overcomes and even vindicates itself against the commandment, 'Thou shall not steal.' In the poverty of wealth we read, too often, the same indictment. '*Crescit amor nummi, quantum ipsa pecunia crescit.*' The love of money increases as much as the money itself increases, until positively the wealthiest man may come to the insane conception that he is absolutely penniless. I have known such a person, and have seen in that person every habit and form of penury assumed up to the positive act of theft for supposed necessity. Insanity, some one will say. Yes, relatively, poverty is insanity under which to steal is an act of forced insanity, aberration from sane law. But both the insanities, the insanity of the starving pauper and of the grasping millionaire, have the same origin in the mental physical life.

The parallel runs further. By the extreme extension of poverty the right to take without lawful possession becomes an idea which covers the conscientious doubt that ought to exist as to the taking. Wealth leads to the same moral blindness. This great estate which the wealthy man owns, near other estates loosely guarded, tempts to extension. This fence, or wall, or landmark, is the fence, or wall, or landmark of the great owner. Why should it remain there? In so great an estate a little more added to it will never be recognised, and that unguarded no-man's land will be better cared for under better protection. Extend, therefore, the fence, the wall, or the landmark. Who shall calculate what theft of wealth, in the very form and pressure and excuse common to poverty, has thus been perpetrated all the world over?

V. I have said that poverty leads to poorness in mental attribute. It leads also to *lowness* of pursuit. The men who live in poverty—and these are my men now—if they must have what is commonly called recreation, seek it in the lowest depths of human activity. To them the forms of sport which are attended with hazard, excitement, cruelty, are the delights of life. They frequent in their miserable way the shooting match for pigeon sport, the race-course, the betting ring, the fancy; and as they cannot go out shooting in accordance with the law, they go out without legal protection, and add the game-keeper, if necessary, to their game-bag. Some, and not a few of them, take to a vagabond life on the sea, get cast off in fashionable foreign watering-places and become the begging outcasts there, revelling in their way in foreign bedevilment.

Wealth follows the same course for its unwholesome recreations. It goes in for sport, for hazard, for the chase, for the betting ring, for the fancy, for the field, for the yachting wandering sea life, for the foreign loitering and polite vagabondage. The parallel is perfect.

I was visiting a neighbouring country. At the hotel I rested at there was staying, at the time, a rich noble, young, handsome, profuse animal to the core, animal from head to foot. He had a retinue like himself at his table. In the yard and stables outside he had another retinue, a ragged, jagged, clothes-torn, flesh-torn crew, who waited for their elegant brethren within to come out and play with them. The game they were both at was badger-hunting. The delight of wealth was to follow the wretched badger into the most secret place it could find for protection from the animals that pursued it. The delight of poverty was to get ahead of wealth in the game, and to show its own animal nature as the best and fiercest. I made inquiry, and if I were to tell the whole story of that wealth and that poverty, I should have to tell a history of poverty and wealth stronger in outline than the first history I have related. I should show that the wealth of that rich animal did nothing and went for nothing else but to feed poverty: to make its owner and all under his fatal ban the possessors of that destitution of mind, body, and estate, which is the very acme of the combined evils of poverty in its most degraded caste.

So true is this aspect of the poverty of wealth, in so many and staring forms does it present itself daily for observation, that, if a man of wealth, for any happy reason, devotes his wealth to some useful and noble purpose, he presents so strange and improbable a picture that he is actually glorified for having done a common and commonplace right thing. He can become, in this way, cheaply, an important and great man, a benefactor, a munificent patron. How he got the wealth will then never be inquired into too nicely. It may have been by fraud, quackery, gambling, sin of any dye. Never mind. He has applied it usefully. He is a great patron. Marvellous! Marvellous! He has held great wealth and has not squandered it. Oh, the princely virtue!

VI. We know of poverty that it leads men to deficiency of self-respect.

'My poverty and not my will consents,' says the poor apothecary.

'I pay thy poverty and not thy will,' retorts the tempter, as he throws in the gold.

How true a picture of the poverty of wealth, and to this day real as in the time when the real story was enacted.

We are certain always, when and wherever a man's poverty is paid and not his will, how bad a bargain and how low a bargain is being struck. Neither receiver nor giver have anything to lose, and so they barter. It is one poor creature helping another poor creature to stagger before the world in an imposing attitude. The mental attributes are in both cases the same to the letter.

> I have money, you have skill,
> Of my money take your fill.
> Of my skill, for your fine gold,
> Take as much as you can hold.
> Thus they barter, nothing loath,
> Till the Devil buys them both.

By which is meant that each of the two high and mighty contracting parties are equally empty and equally ready to be easily caught in the devil's net.

The most debasing thing of all things human is, in fact, poverty selling its soul. No, there is one thing worse, and that is wealth making the purchase. Yet even in the present day, this low and heathenish thing, this barter of the highest gift of man, is one of the common stocks of the market-place, and one of the commonest causes of abject poverty.

And a further evil in this transfer of soul for money, this barter between poverty and wealth is, that vanity is fed by both. There is a vanity of poverty as of wealth; a pretence, on each side, of humility by the poor, of condescension by the rich, which goes to the root of good morals and tears the root up, root and branch. The obsequious bend of body is reflected in the rich man's complaisant nod; the low curtsy is the mother of the great lady's sweet smile. Both sets of acts supply mere food to vanity, and are both wrong, because both are false. In the one there is no real respect, in the other no real feeling, no sweetness; but barter, backwards and forwards. The nod, the sweet smile, are bought as much as the bow and the curtsy; each is the automatic declaration of poverty—the one of the soul, the other of the body.

VII. The effect of poverty is to lead to bad training of the young. Why should the young be trained to anything more than their like? Who shall take care of a child that cannot take care of itself? Who shall take up the trouble of parentage when there are public institutions, from the workhouse upwards, ready to do the work? So the young poor of the lowest class are left to their fate: they go their own way, and unless caught by some intermediate hand in their earliest and most impressionable days, and put to useful, honest work, they follow in their parents' career, like breeding like, and the child becoming the father of the man. The same extends to wealth. After the child of the poorest, there is no object of humanity, as a rule, more neglected than the child of the richest. Thus it has become almost a byeword to look upon the children of the immensely rich as the ne'er-do-weels of the community. We were trying, a short time ago, at a literary table to discover if there were any or many instances of a man enormously rich leaving a child of any mark of character or name. We could find plenty of illustrations on the other side: plenty of instances of poverty of character begotten of riches, but not one example of nobility of nature so begotten. 'It is,' as the wise man said, 'a sore evil; namely, riches kept for the owners thereof to their own hurt. But those riches perish by evil travail; and he begetteth a son, and there is nothing in his hand.'

VIII. So far I have traced the poverty of wealth by two parallels of what may be designated the moral side of the question. I reserve to the last the purely physical side, that which relates to failure of mental and bodily health, to disease, and to reduced value of life by poverty and wealth alike.

The mental qualities of the untrained illiterate poor and the untrained illiterate wealthy are so alike, that but for surroundings it were impossible to tell the difference. Each set entertain similar superstitions, fancies, and follies. In sickness they are caught by the same sounds, the same pretensions, the same hopes, the same fears. They require to be directed as to what they shall do, without any exercise of their own judgment or discretion, and they accept the value of professional advice to the letter so long as they have faith in it. But as the trust is on faith alone, it is of little worth, the merest change or caprice having the tendency to upset it altogether. For this reason they are ready in illness to run to any man who, as they have been told, has effected a cure in the same case as their own. The professed and open charlatan lives, in fact, on the very poor and the very rich. I have for some years past presided over a society that has done an immense deal for clearing the moral atmosphere of some of the worst characters of the outrageous charlatan fraternity; and the conclusion I have been obliged to come to, from the direct observation of what I have observed, is that these harpies have no classes of society upon whom they can firmly rely except the two that are most separated—the lowest poor and the leisured wealthy. The poor are the decoys of these persons; the rich are the dupes. This is probably the case in regard to all professions; it is certainly so in regard to that profession which has to treat disease;

and what the luxuriously wealthy, as well as the poor, will believe from the charlatan, what they will undergo, surpasses any descriptive teaching I have either time or inclination for setting forth.

The two classes resemble each other in the nature of many diseases from particular causes under which they suffer. We are constantly speaking of the poverty-stricken as an intemperate class. Sometimes we hear them pitied in regard to intemperance, sometimes blamed. 'Give them good and comfortable homes,' says one man, 'and they would not be intemperate.' 'Take from them the drink,' says another, 'and the good homes would follow as a matter of course.' 1 do not discuss this question here, because I am treating of results, and the results are, without any doubt, a large mortality, preceded by an unnecessary amount of sickness and pain, from drink as the cause. These poor resort to drink as a shroud from their misery. 'Let us eat and drink, for to-morrow we die.'

Amongst the leisured and luxurious rich the same rule obtains so closely that we may, as it were, step over the great intermediate classes, from one class to another, just as in a large hospital we may walk across a corridor from one ward to another. The total of disease from strong drink amongst the wealthy unemployed, appearing under all forms of cunningly devised fables as to its nature and name, would require a volume of no mean size to interpret honestly; a course of lectures, not a sentence or two in one essay. Education, warning, proofs of danger of the most practical kind, seem to have no effect. Quite recently I have had brought before me the very painful picture of four youths, all of highest class in regard to education, killed by wealth, three actually, one morally, all from this cause, and all within a period of six years. Here education utterly failed. But not long since I saw wealth reduce to poverty of death the last of seven members of a family who had received no education. These were competent, frugal, comparatively happy working people, living in sufficiency by the work of their hands. Suddenly they get what is styled a windfall, but which I would rather call a whirlwind fall. They wake one morning to find themselves rich. They must, after the bad fashion of the bad wealthy, signalise the event by a family banquet, with friends invited from all quarters. It must be a grand affair. They must, for the first time in their lives, taste rich wines, costly withal; manufactured champagnes, bearing cabalistic names, made at tenpence a bottle and sold at ten shillings. Now the poverty of wealth begins. 'Let us eat and drink, for to-morrow we die.' And the whole proverb is realised to the letter. I have seen that unfortunate family die off from the poverty of intemperance, enkindled by wealth, until not one remains of all who laid the table immediately after the whirlwind of their fortune and their mortality.

From poverty and from wealth there occur degenerations of the organs of the body which lead to loss of physical power, feebleness of mental power, ready lassitude, indifference as to external surroundings, carelessness as to the sufferings of others, and sentiments for self which absorb all other. These qualities of mind and body, although looked upon as faults, are really and truly results of organic failures dependent upon degene-

THE POVERTY OF WEALTH 249

rative modifications of the vital structures. It is a part of the healthy bodily life that the food, or force-making substance, that is put into it shall produce, or yield, or give out all that is put in, in the form of work done by the body. If sufficient be not supplied, then follows the change or degeneration due to want; if too much be put in, then follows oppression, laying by of matter that cannot be used, and the same kind of enfeeblement as that which springs from want. Between the action of a heart under-nourished and of a heart overburthened, there is practically no distinction of impairment, and so we get similar phenomena from what seem to be quite opposing causes.

Under both circumstances we obtain shortened life from impaired function. The muscles lose their activity, their willingness to work, their readiness for labour, their powers of endurance. The brain, in like manner, grows dull, and the mind inadequate for duty. Thus arises that misery which marks the face of common poverty, and that melancholy which so often settles in the expression of the wealthy poor. The expression tells the common story of the unfavourable course of the illnesses from which they suffer. Of all the sick, they are amongst the saddest and are most sensitive of danger. Of all the mentally afflicted, they are amongst the most prone to melancholy.

And from this source of degeneration follows, in certain turn, that failure of heredity upon which the stamina of nations becomes so faulty. The children of wealth, the golden children, as the sycophants to wealth designate them, are, as the children of the poor, the frequent inheritors of the feeblenesses of their progenitors, until at last they fall into the same category and, by the rule of the decadence of the unfittest, speedily join the veritable ranks of the poverty-stricken, to sink with them. We have in the world now thousands, perhaps millions, of poor who are by this reason the last remaining links of once great and powerful, because healthy families, whom wealth had not yet blasted, or whose fragments poverty had not yet attracted unto itself. By careful research we might find in every large workhouse in this kingdom the proof direct of this natural law of denudation of wealth. The geologist's denuded mountain is not more distinctive, and is certainly not a more common or striking phenomenon.

Perchance the reader, recognising the comparisons and contrasts which I have drawn, will see also the current dangers incident to the extremes of poverty and wealth, and will realise the fact that wealth and poverty in the same sphere are like two electric clouds, safe enough so long as a friendly and sufficiently wide distance separates them, but certain as death to explode in revolution if they too nearly approach to each other. Let the poverty of Eastern London come too near the wealth of Western London. Let some moral disturbance, some panic respecting supplies of food, some want of work, disturb the equilibrium of wealth, and a mightier consternation than Imperial Rome herself ever witnessed were easily at our doors.*

True! But some one may exclaim, 'Tell us, O physician of health, what is the remedy. You have laid

---

* Since this address was publicly delivered on January 16th, 1886, we have had a first taste of this danger in the West End riots.

out the diagnosis, you have detailed the symptoms, you have declared the danger; tell us the mode of prevention.' There is a theory on this subject now very widely spreading, which I cannot endorse. This theory in various guises is covered by the term Socialistic Development, or Socialism. Mr. Fox Bourne, in his late article in the 'Gentleman's Magazine,' has most ably explained all the four quarters of this ideal reforming method. He has described the Scientific form of it, the Anarchic, the Christian, and the Æsthetic; and he has laid bare, with almost poetic fervour, the hopes and expectations of multitudes who see in their dreams of the future the millennium of their desires.

There is much in the picture which he has drawn with which I sincerely sympathise; there is much which, in my outline of a model health land, I have myself conceived and proposed. But with the mode of reaching what is yearned for in the socialistic propagandas of all quarters there is presented to my mind that very poverty of wealth which ought of all things to be most avoided. It is as though the propounders of these theories, having their gaze fixed on the poor wealthy creatures, were striving after the self-same poor wealthiness, or self-same luxury of indolence. Take, for example, an illustration. It is made an argument that if all labour were equally divided, two hours per day is all the labour that need be thrown on any man, woman, or child; so that out of twenty-four hours, if eight went for sleeping, two for feeding, and two for working, there would be twelve hours left for pleasure, or for such work as should exactly please the owner of the happy time. This division of time is just about that which the wealthy man obtains, and what is the result ? It is not health, it is not strength, it is not happiness; it is *ennui*, despair, poverty of mind, body, and life, in which the mind is no longer a kingdom, but a prisoner in a golden castle.

The error might be pointed out as extending to all the mischiefs personal and general of wealth, to feeding and feasting, to bad sports, to low habits, to indifference, to the whole moral of the practice, ' Let us eat and drink, for to-morrow we die.'

No! in this crave for happiness by the luxury of wealth I, for my part, see no millennium. I cannot see that we are constructed for any such millennium. As long as we have to live by eating we must live by working: we are as yet too gross to subsist on ethereal nothings as the angels do. So every violent measure for passing from the poverty of poverty to the poverty of wealth is certain failure.

The reformation, as I venture to argue, is the same as that which Joseph Mazzini applied to the perfection of democracy. It is 'an educational problem ; it is the eternal problem of human nature,' in which the idea of wealthy leisure and miserly rest plays no part whatsoever. The true reforming method is to instil into the minds of all men and all women the correct relationships of wealth and poverty ; to exalt labour as the foundation of health and wealth ; to expel the idea of rest on wealth ; to teach the emptiness of the fallacy that it is either good or reasonable to provide for generations of unborn idleness ; to show that all attempt to base the continuance of family name and fame on worldly possessions, mapped out from the dead earth by so much poor parchment covered with legal hieroglyphics, is but the insanity of vanity; and to impress on all minds, but on the minds of the young especially, the vital truths: that the family which shall live longest in its units, and in its membership, is the family that leaves the healthiest progeny: and that the nation which shall live longest is the nation which, being healthiest, is by necessity strongest, most active, and nearest to that eternal energy which would itself be dead if the atoms it animates were to fall into repose.

UPPER AND LOWER LONDON.

A PROJECT FOR RADICAL RECONSTRUCTION.*

A CLEAR sky in a place like London; clearer and cleaner streets; the emancipation in close places of one person from the empoisoned breaths and emanations of another; room to breathe. These are the blessings the people are looking for from their sanitary deliverers.

NEW UPPER LONDON.

If I were going to be one of the competitors for the munificent Westgarth prize on the purification of London, through the reconstruction of it, I should propose such a reconstruction as would be, in an architectural sense, perfectly revolutionary. I should propose to go upstairs to the tops of the houses to do it, and should indicate the construction of a new London, overhead, Upper London.

To convey what I mean, let us move to the best constructed, as well as the most beautiful street of this metropolis, if not of the world—Regent Street—in the part called the Crescent. That is laid out for such a design as if it had been prepared for the experiment. All the houses are of the same height, and the height

---

* From an Address delivered before the Society of Arts on Wednesday, March 26th, 1884.

throughout is just right for a city like ours. It is sufficient to be handsome and commodious without being overwhelming, and without excluding the light from the streets. The roofs of Regent-street, at this part, are flat in comparison with other roofs. They are utilised here and there by the photographers for their studios, which, although temporary structures, stand firmly and well, in ready communication with the houses on which they are placed. The studio, where it exists, seems, naturally, to form and become a part of the house.

When we glance along the line of roofs, as on a level terrace, the idea of reconstruction of all roofs, and of the re-adaptation of them, becomes very distinct and suggestive. The width of most London houses averages, as near as I can estimate, about twenty-five feet from front to rear. Here, then, is good space for a terrace for foot-passage. Imagine along two lines of long streets a terrace of this kind, with a handsome railing on each side, a perfectly level floor surface of wood, and, at intervals, light bridges spanning from one terrace to another, and you have an upper-day London which might almost relieve all the pressure from foot traffic in the streets below. Each house would have its own exit, or door, at the upper as well as at the lower parts; and at convenient spaces, each terrace would be accessible from the street, as the Holborn Viaduct is at the present time.

It suggests, at first, a revolution of ideas to conceive such a change. It suggests much out of which a humourist can for a moment make capital. I know all this very well. But there is, in point of fact, nothing more in it than in the first idea of making a tunnel under the streets or under a river. When the suggestion is looked at bit by bit, without prejudice, it offers more of sanitary advantage for the purification of the atmosphere, the protection of property, the comfort of the people in transit, the lodging of the people, the exercise of the young, and the beautifying of the whole city, than could be entertained on a mere general statement of the proposition.

In the first place, for every house in connection with an upper terrace there would be the most perfect through and through ventilation of air. The staircase would no longer be a closed cupola for holding and storing all the emanations from the basement upwards.

In the second place, the fact of having terraces on the upper surface of London would lead to immediate arrangements for the purification of the air from smoke. So soon as the roofage was accessible as a terrace, the plan which Mr. (now Sir) Spencer Wells projected for the removal of smoke from every house, by laying down horizontal conducting tubes with central exits and smoke-consuming furnaces would be easily practicable,—presuming always that some smokeless fire be not invented, or that coal gas do not become the fuel of the people. These terraces would then be the healthiest parts of London; charged with flowers and trailing evergreens, they would be the empyrean gardens of the great city.

The terraces, with their light intercommunicating cross-bridges in the long thoroughfares, would be more than pleasant footways and shady lanes for the foot travellers, or for travellers in light noiseless vehicles, like tricycles; they would be most useful for other purposes. Along them the electric lines would pass and enter the houses direct; and from them the letter-carriers would most easily deliver their letters.

These terraces, while relieving the traffic in the streets below, would remove all necessity for the fire-engine, and would make London practically safe from fire. From them water would be supplied readily, a trained police for this upper London being ready at every moment to go down and extinguish fire in every domicile, carrying the hose with them, or plying it from above.

I think that no one who reflects will fail to see that all these changes would be advancements of great value for the health of a city like ours. They are, however, not the chiefest advantages. If any one will take the trouble to go, observingly, through the busy parts of London, where there are long miles of roadway—along Whitechapel and Mile-end, for example—he will see the most jagged, hideous lines of roofage. Here a line of houses two stories high; there a row of three or four stories; then a single house of five or six stories; and so on, over and over again, like a set of bad teeth.

If the plan here suggested were carried out, all this would be rectified. A street like Regent-street expanded into a straight line would extend from the Marble-arch to the City, and from the City to the extreme East-end. The line of terrace pitched at five stories would necessitate the building up to the same level all the houses in that line, by which at least one-fourth more houseage would be supplied, with arrangements for giving comfortable and healthy homes, beyond what now exist, to a fourth of the present population. The suspension cross-bridges would not be without their compound service. They would be the bearers of electric lines along their side-ways, and would probably soon be utilised as centres from which electric beacons would be suspended to light the streets beneath.

Imagine the metropolis turned into a fairy land by this adventure of science into the domain of art, and art reciprocating the idea, with all her rich resources, and we see in our mind's eye what our children, when we are all of us gone, may really see, and perhaps thank us for proposing for their benefit.

Objections may be made about mechanical and architectural difficulties. I heard them all made when the Holborn Viaduct was projected; I saw them all melt away as Colonel Haywood's practical mind came into work, and as his unthanked skill and industry, and responsibility and genius carried all before him.

It will be objected that flat-roofed houses are not weather-tight. In the year 1825, the then Parisian Asphalte Company roofed two houses with asphalte in Hinde Street, Manchester Square. I lived in one of those houses for twenty-eight years, and a better roof I never knew; but for the London smoke it would have been made into a garden. Men working upon it, walking over it, communicated no sound whatever into the rooms immediately below.

It will be objected that houses will not bear the weight of superimposed suspended terraces for foot-walks. If they will not, they ought. In no direction would the sanitary improvement for the purification of a great city be more useful—as a side improvement—than in so reconstructing defective houses as to make them capable of bearing an equalised weight, which, carried by many, would, as we know from the bearing of ice, be comparatively light and practicable.

It is not to be disputed that in many instances dilapidated houses would have to be razed to the ground for the purpose of obtaining sound and safe foundations, and to those who maintain house properties in a condition requiring such an extreme measure, the idea suggested in this paper would, naturally enough, be objectionable. With the artist and sanitarian such an objection could not, however, carry much weight, because we are all agreed that for the health and beauty of the metropolis no reform is so much needed as that one which should be based on the demolition of old and dangerous tenements. In the course of such demolition and reconstruction, owners of property would themselves, in due course, be amongst the most fortunate, as sharers in the improvements.

Of the many plans which have been suggested for giving space to crowded cities, such as terraces in the streets opposite to first-floor windows, or tunnels subterranean, none seem to me to be half so practical, half so likely to secure the purification of the atmosphere as this, which I have now for the first time, after some years' hesitation, ventured to sketch out, not as expecting ever to see such a project realised in my own time, but foreseeing the necessity and applicability in times to come.

The project here described, to which Mr. Westgarth has, in more than one paper, given his powerful commendation, has taken well with large numbers of the community, and many an inquiry is made of me when the pleasant and, as most people admit, easy reform is about to be set on foot. I answer always, *wait*, and little by little it is sure not only to be commenced, but to be carried out to the fullest extent. For the day is not far distant when, from sheer necessity, traffic through the streets of London must be relieved by the introduction of some new plan; and, when that necessity comes under discussion, then the Upper-London project will naturally force, and probably soon win, its way.

New Under London.

While Upper London is waiting there lies before us another, and, indeed, more pressing need, having relation to the London under the feet of Londoners, and having reference to the regular and effective removal of the most dangerous refuse of the household, with the utilisation of that refuse for the benefit of the human family.

What millions of money have been spent in the effort to purify Under London we all know. What have been the results of that expenditure—not to use the word success—few know. Some of us, of the Society of Arts, at the instance of Mr. Edwin Chadwick, whose name as a sanitarian is a name of the century, did, some time ago, commence an inquiry bearing upon the important point now under consideration. We opened, or, more correctly speaking, we reopened, the question practically, and we discerned all at once—although our inquiries were entirely confined to limited areas of London —so much evil, that we rather abruptly closed the evil up again as one we were not prepared to remove. But sewage, like murder, will out, and enquiry must proceed.

What we did discover was in truth so serious, that the wonder we laboured under was how London can be so healthy as it is. We found that London is still honeycombed with what are not in name cesspools, but are, in fact, sewers of deposits; a truth we all practically recognise by the second-hand measures we take to meet the primary blunder. For example: from the window at which these remarks are written I see that one of my neighbours, the owner of a large house, has carried out of his house, from the basement, a three-inch tube far above the level of the parapet, in order to deliver into the air any gases that may accumulate in the main drain of his residence.

It is not good for the air which I and many others have to live upon, that it should receive the foul gas from the decomposition of my neighbour's organic excreta; and if everybody's neighbours did what he has done, it would be detected, in some weathers, that the process must be stopped, as it was in a former day on a recommendation of the Royal College of Physicians. I do not, however, blame my neighbour for what he has done, because I have done it myself. It is a natural method of self-protection amongst those who know best how to protect themselves. My contention is, that the necessity for any such method is proof demonstrative of the rottenness of the primary system which causes the necessity, and which keeping us foul beneath our houses, makes the air, at its best, foul also above them. My contention is, that the decomposition which gives origin to the gases that are let out by thousands upon thousands of channels—by tubes from houses, by soil-pipes within houses, by accidental openings and pores in all directions, by gullies in streets, and by great outlets of sewers from accumulations of sewage—ought never to have been permitted at all. My contention is that the sewage, removed clean away, unintermittingly, many miles from the community, without having decomposed either above or below the living-place, need never pollute any place nor have any destination except the land which is calling for it, and which dies if its demands be not naturally supplied.

The complete removal, at every moment, of all the organic excreta is, however, still an unsolved difficulty, which, remaining unsolved, is a block to every step towards perfect purification.

Combined and Separate Sewerage.

We are yet distracted with the debate ever going on between the advocates of the combined and the separate system of drainage. Shall all our organic excreta go with the storm-water into the river and sea—*the combined system*; or shall the water go to the river and sea, the sewage to the land—*the separate system*?

Unlike our neighbours on the other side of the Channel, we have agreed to give up the cesspool, and have not agreed, as they have, that all disposal of sewage in running streams shall be legally prohibited. But, in giving up the cesspool, have we greatly advanced, so long as we pollute the running stream and lose the natural fertiliser of the land? Looking back on all the controversy for the last thirty years, and reading back still further, I feel we have not advanced. I do not think it would be wise to return to the most scientific system of cesspoolage, but I cannot conceive any next worse plan than the plan of passing the sewage with storm-water, even on the most scientific system, into running streams, and of robbing the land of its greatest requirement, the fertilising treasure which is so systematically and recklessly cast away from day to day.

In London the combined system having obtained most favour, and the people having without stint poured their wealth, as well as their sewage, into it, the system has had its best days and its best chances. Everything has been said and done for it that could be said and done for it. Time has been given to the trial of it with a patience unexampled. The advocates of it, the sustainers of it—and let all credit be given to them in these respects—have laboured for it with a perseverance, a skill, and an enthusiasm worthy of a dozen better causes. But what has been their success? Let the great Outfall failure tell its own story. It will be most interesting at the close of each year to see what has been the expense for outside deodorisation alone in twelve months. What has been the cost for the material alone of the thousands upon thousands of tons of so-called disinfectants poured out to purify what ought never to have demanded purification! Mr. Chadwick says 200,000*l.*; and every penny of this cost a penny's worth of proof of the radical mistake which has been committed, and which, by the natural force of example, has been extended from the metropolis into other great centres.

Since the gigantic difficulties in respect to the great outfall, the pollution of the river at the outfall, and the loss of the most valuable product which London supplies, have become fully recognised truths, the people generally are asking what is to be done. Then comes the scare of approaching cholera, and loud outcry and busy fussation of tentative schemes, and serious complaints of heavy tentative expenditures, with a happy subsidence of the scare, followed by an unhappy subsidence of complaint. A period, in short, of waiting for another scare, and of waiting longer still for the verification of the *bona fides* of the scare in a grand epidemic and a hasty resolution to make some change for better or for worse.

The true sanitarian does not acknowledge the soundness of this principle of waiting to be forced to do. He wishes the best thing to be done at the best time, that is to say, at the time when the mind of the community is free from all excitement, and when the mind of men of science is open to work without pressure, and with the feeling of perfect freedom from all haste and worry. The work that is required to be done should be done now while there is pause; and it should be done so effectively, that it will not require to be recast for a century at least, unless some new and radical discovery should be advanced that will provide a radically better principle and practice. If our engineers had kept to their own splendid department of science, instead of assuming the position of doctors of health, or if the doctors had been stauncher at the first, all the present trouble might have been saved. By reverting to their true positions and functions, both can still make a magnificent amend.

The Work to be Done.

What it is requisite now to do for Under London is so obvious, and, in fact, so simple, that it may be told in a few minutes. If I am correct, moreover, in what I have learned, the prime cost of the doing would be comparatively small, while the returns would rapidly repay all the original outlay.

The first thing to be done is to come direct to the natural and only common-sense principle, the separate system of disposal of the sewage. '*The sewage to the land, the storm-water to the river,*' is a proverb which every boy and girl at school should learn, and every man and woman who has influence should enforce. It is the basis of all true and safe sanitation. It is obviously the method of nature herself, when she is left to her own devices : and where could we find a better teacher ? If, indeed, her scheme for this purpose were not so grand that our worst ignorances cannot thwart it, all vital motion on the earth would cease, for it is her plan, out of the very womb of death to bring forth the most perfect forms of life. She builds up the earth and all that is earthy by primary burial of the earthy; she vitalises by the living waters moved by the sun and the moon. Were she to drop all her treasures into the sea, as some of her children do, there would soon be no land on which things of life could live.

We Londoners, therefore, must, if we would do the best thing in the best way and for the best purpose, follow our mistress and guide, in her mighty cleansings of the dead from the living.

When we have once got this principle in our minds and are brought to the application of it, we may look at what it is we have got already, before we determine what it is we have to change. Fortunately, we may discover by this inquiry that what we have got, faulty as it is, admits of being utilised, to which utilisation we should direct all our efforts.

A system of sewers within the present sewers is the project we ought to insist on; of sewers limited and perfect for conveying from every house its own sewage together with the water used in the house for its own purposes—no more, no less.

And for the storm-water, a system for its escape as storm-water into the river through the existing channels, outside the true sewers, and absolutely separate and distinct from them.

Towards the attainment of these objects there is no difficulty that cannot be overcome. All is laid out for the work. Whether the inner sewers should run along the floors of the existing sewers, or be raised above those floors, or be suspended from their roofs, is a point of detail that may have to be studied and experimented on in order to prove which position is most convenient. This detail does not affect the method.

The material for the construction of the inner or true sewer might probably be settled on at once, because on this matter there is experience for a guide. In some American cities, and in some English places, iron tubes are found effective, and nothing could be better here. Such tubes are already constructed for conveyance of water and coal gas; they can be perfectly closed throughout, so that no sewer gas need escape from them, and they can most readily be made to communicate with the main drain of each house.

The size of the true sewer may be a subject for consideration, but the question can easily be solved from existing data. My own view would be for small sewers, a four-inch drain for a house of twenty rooms, with the sewer in proportion, as Mr. Chadwick has suggested, and planned according to the oval shape which he also has described so convincingly.

Taken systematically from one district to another, London might, in this manner, be sewered throughout, without the change being known by the masses of its great community. There need be no opening of streets to lay down new lines of sewers, no important stoppages of existing outflows from the houses. As the inner sewers were laid down within the existing ones, the connection from house to house could be laid until all was complete. Moreover, the connection between one house and another, now so very imperfect, could, during the change, be rendered absolutely perfect.

As a method of conveyance of the sewage of every house away to some common escape from the town or city altogether, this plan admits of the application of yet another advantage. By the addition of an exhaust pumping system at various sumps or stations for bringing the sewage freely and regularly out of the houses from which it flows and on to its final destinations, the houses themselves, ventilated with the best obtainable outside air, might have the air carried away through every outlet from them into the central sewer. The water-closet, instead of being charged with the emanations from the closet, would be flooded with air from without, which air would flow down the closet, and away by the main drain, as the water from the closet flows into the drain. The same with the sinks and every other of the openings, where there is an exit for the contents from the dwelling. The bad air from slums, mews, shambles, vegetable markets, dust-bins, and all repositories of objectionable matters, solid, fluid, aerial, could also be carried off by properly constructed communications through the exhausting central sewer.

The Destination of the Sewage.

It will, I doubt not, be admitted freely that the attainment of such a system as is above contemplated would be a result satisfactory beyond expectation. But what, it will be asked, is to be the destination of the sewage flowing through the central sewer, and what of the storm-water flowing by it, but unmixed with it?

Let us take the disposal of the sewage first, as the most important and difficult question.

It will be seen at once that the enormous difficulties at present existing would be exceedingly reduced under the system above proposed, by two changes—the tremendous reduction in the volume of material that has to be disposed of, and the condensed character of the sewerage itself. We should be dealing, not with diluted decomposing material, but with a substance undecomposed and ready for application, as required for purposes of cultivation. This reduction in volume, which might be still further promoted by a reduction of the water supply to the extent of ninety million of gallons per day, would render the fluid manageable both for collection and transportation.

The plan I suggest for direct disposal of the separated fluid is that there should be constructed at, or near to, the present outfall, or at any better place on the river, a series of vast covered docks or reservoirs, into which the fluid from the prime sewer should be pumped, and in which it should be retained for brief periods until its removal. To effect removal from the reservoirs, I should suggest floating barges or tanks, which, loaded with the product, should steam away with it and carry it wherever it was destined to be borne—to the sea, if it must needs be thrown adrift; or to some point of land for utilisation.

Whenever it was made known that material of such value for the land, as the contents of these floating galleons of wealth, was in the market, we cannot but suppose that the demand for it would be instant. There are men living who can remember the time when the Bristol barrel filled with the same kind of product went from the Belgravian fields to the West Indies, and—tell it not in our sanitarian Gath—came back loaded with the same product transmuted, in the laboratory of nature, into sugar! Why not in a better form, in these better days, transmit the same kind of substance to different parts of the earth, or to barren portions of our own islands, and there let nature transmute it into fruit, sugar, or other useful substance, in such a way that it shall administer its full benefits to mankind? Barren portions of our sea-coasts could, by these modifications of the separate system, be made the most fertile and beautiful of all our tracts of vegetation.

To the engineer, when once a system were decided on and declared, these modes of transit, with many improvements, would speedily occur. With the engineers it is not my province to interfere. They exist to carry out what has been determined on, and when they know what the people want, they will do what is wanted, as surely as they will lay down, after the country has said they must, a new railway or a telegraph. We have but to declare the principle, and get it declared, that every town in England must be cleansed of its organic excreta out and out, day by day, as certainly as it is supplied with the food brought into it, and the thing were done.

Toward such perfection any powerful movement steadily and resolutely devoting itself would soon be backed up by the common sense of the people, who require nothing more than a competent authority to settle the subject determinately. The utter failure of the combined system as a permanent solution of the drainage difficulty, and as a mere transition from the cesspool to the method of separate removal with immediate and fruitful utilisation, is becoming apparent with such swift conviction, that there will soon be an outcry, unless, irrespective of all interests except the true ones, our authorities will undertake the necessary radical reformation.

The moment we can, by the skill of our engineers, get our in- and out-going drainage system as good as our railway system, as true in its working, as continuous and systematic, the most important of all our basic sanitary reforms will have been introduced; while until this basic reform is carried out there can be no sound sanitation at all. I have asked many times, sought many times, for so much as one instance in which the combined system of sewerage, apart altogether from its loss and extravagance,—which might be tolerated if its results were good—has proved a sanitary success. I can find no instance of the kind.

The towns which depend, as London does, on stormwater to flush and inundate their sewers, are like the old mariners who depended on favourable weather for a favourable voyage. The day has passed for that hazard.

The last difficulty that has to be met relates to the disposal of the storm-water, and that is solved, or may be solved, in the one sentence, '*the water to the river.*' In some places, where the river is small, it might be well to let the proverb have its full play; in London it might or might not, for the Thames is large enough either to receive the storm-water or to do without it. In London, consequently, we might catch the storm-water, if that were desirable, in separate reservoirs, and use it for laundry work and many other valuable purposes.

Thus on all sides the advantage of the separate system would turn to good account in the mighty Babylon which we have reared to-day. By it, our Babylon, in its worst quarters, would be ventilated; would at every minute of its life be scoured, cleansed, and purged; would send forth its dead and dangerous refuse for living and useful purposes; and would insure to itself good and pure water, for that cleanliness which is next to godliness, and without which godliness itself were a rare virtue.

Upper London a garden of terraces and flowers, beneath a smokeless canopy. Under London a double river, one of dead wealth, the other of living water. This would indeed be London.

'FELICITY AS A SANITARY RESEARCH.'[1]

Our Congress this year has been ringing peals of congratulations, and the peals are deserved. Sanitary science, in some simple directions, has won triumphs such as have never been won before, and its advocates, once called enthusiasts, dreamers, visionaries, and other poetical names, are now, in respect to enthusiasms, dreams, visions, looked upon as commonplace observers. The miracles they declared possible are performed so regularly that wonder and doubt have ceased.

We sanitarians declared that it would be a comparatively easy task to find out the courses and, at least, the proximate causes of the great pestilences, and that, with a fair knowledge on these subjects, combined with a comparative ready assistance by the public, we could control both the courses and the causes. The work has not been done so thoroughly as we could have wished, because the public has not come up, as yet, to our views; but the work is progressing, and sufficient is already done to prove the truth of our position.

We said that the once current rates of mortality were deathly of deathly; that they represented a low civilisation; that they might, in these countries, be

[1] An address delivered to the Congress of the Sanitary Institute, held at Glasgow, on Thursday evening, September 27, 1883.

reduced generally to a mean of fifteen in the thousand per annum, and, in favoured localities, to a lower figure still. We were taunted with the rejoinder that if such were accomplished men and women would live a hundred and fifty years. We replied, Let them live two hundred years if they like, but let us, any way, reduce the huge mortalities which are considered natural.

The result of our work is that there are towns where the average mortality is actually lower than fifteen in the thousand. Strangely, too, the popular cry is not now against us as enthusiasts, but against towns which do not follow up our enthusiasm. Towns, therefore, in this day, compete with towns for a low death-rate, knowing well that should mortality from temporary and as yet accidental causes rise to what was considered the mere natural a quarter of a century ago, they are temporarily ruined if they depend upon outside popular favour for existence.

All this is most satisfactory, and would afford a fruitful theme of discourse. But what I want to-night to rivet on the memory is a new thought for new work. I want to put the following few questions and to endeavour to answer them.

Can we honestly believe that these triumphs of ours, which have so far ended in a certain victory over death, have introduced any fraction of triumph over misery?

Have we by our labours assisted to make men, women, and children happier as well as longer-lived?

Have we tried to effect anything in that way; or have we, aiming at nothing more than the promotion of a longer life, left the rest to chance, as if it were not our duty to include human felicity in our design of labour?

T

.Can we effect anything to ensure felicity as well as length of days? In other words, is felicity a subject open to sanitary research, and if so, in what directions shall we labour for it?

These questions are momentous, because if we are aiding in the art of adding to length of life and in developing population without giving to an extended and universal life felicity, or the enjoyment of that which is given, we may in the long run be working evil rather than good for the human race.

A race unhappy lives too long to live.

Surveying the questions I have submitted, I do not think that we have, so far, done anything to add to human felicity. In the first stages of our labours that, indeed, were impossible. We have had to deal with very unpopular, and some think unsavoury, subjects. We have had to be excessively personal. We have been obliged to tell people to be clean both at home and abroad. We have been forced to be fault-finders all round. We have even had to frighten the masses—and fear is a terrible foe to felicity—both in the house and out of it; and until I, one day, ventured to show, by an allegory, a pleasant side-station on our steep and narrow road, we seemed to conceal the destinies we had in view, or to leave them for anybody to discover, an almost hopeless leaving.

We have not tried, therefore, in any direct manner to teach the way to felicity. We may, like the rest of the world, have spoken of health and happiness. We may have commented on the sound mind in the sound body; but we have not tried to systematise effort towards the attainment of felicity as we have towards the attainment of length of days.

I may make these admissions without the slightest compunction or regret. If we have done no good in the direction I have referred to we have certainly done no wrong. There has not been sufficient time for a development of infelicity from extension of life, so we need not ourselves be unhappy.

But now comes the last question.

Can we by any future effort advance human felicity by a scientific research into the sources of it and the impediments to it? Can we, scientifically, connect health with happiness? If we cannot we had better never have been born. We are like preachers of mercy who are empty of charity. We are mere sounding brass and tinkling cymbal.

I do not, for my part, believe that we are in this plight. I hope it is all right that we were born. I believe that we have the moral as well as the physical health in plain subject before us for study, and that we, of all men, ought to see how to combine the physical with the moral, and to understand the relationships of the one to the other and the interdependence of the one on the other. I trust, therefore, that from this present Glasgow Congress we may take a new departure by inaugurating a new school of sanitary students and scholars, whose interests it shall be to learn the physical and moral art of living well, so that bodily health shall, of a truth, be mental felicity. A dream, do I hear it said? If I do, I hesitate not. It was a dream a quarter of a century ago that men could touch death-rates and reduce them to order. To-day the dream is a fact.

I see no reason why we should not, by patient research, know all that pertains to our own lower natures at least. I can feel the astronomer overwhelmed with the sublime story that lies before him. I can see the metaphysical philosopher overwhelmed as he questions the illimitable destinies and the sources of illimitable power and will and being. But we sanitarians, dealing with secondary phenomena; with phenomena repeating at every moment; with our timepiece selves going through regular courses of eating, drinking, breathing, thinking, working, wearing, sleeping; with mechanism that can be counted, measured, weighed, and calculated on commercial values; we surely have no insurmountable difficulties to get over in determining what are the conditions under which human felicity is possible, and what are the conditions which prevent the accomplishment of felicity. The hardness of the task lies, at the outset, in getting a good view of the actual meaning of felicity.

Felicity is contrast to misery. To many minds it is no more than a fancy, an invisible breath of some poet who writes what he has never known, and has never expected to know; or a laugh of some cynic who, tired of life and its vanities, declares that 'all things are alike to all,' and that man, with the beasts, neither goes upwards nor downwards, but dies.

To men who, like myself, are engaged for ten to eleven months each year listening to the sorrows which the sick are forced to tell and the healthy are forced to confirm, it would not be difficult to conclude that there is in the human world no such thing as felicity. There was a noted physician I once knew well, who told me at the close of a career of fifty years of active practice, during which few men had seen more of his fellow-men or had observed more keenly, that he never had met a perfectly or indeed a comparatively happy human being. His view was not altogether peculiar. The professors of medicine generally are felt by many to be stoically indifferent to sorrow, as compared with other persons. They are not so at heart, but, knowing the smallness of human felicity, they are less oppressed than others by the extreme and tenderly acute occurrences of sorrow. They read the Book of Wisdom every day from nature, as the Chaldaic writer of it did ; and so, in the everlasting presence of nature, they become possessed of a demeanour which seems to separate them from the individual life ; and as in that presence felicity is not the feature they are most wont to recognise, they give it wings to fly.

Proofs of Felicity.

By the hard and fast scientific mode of looking at the phenomena of nature, it might seem, then, at first view, that human felicity had no proofs of existence. There are, fortunately, other evidences which give positive proofs in characters as purely scientific as any in the observation of science. Granting that these are exceptional evidences, they are still in proof.

I notice four of these evidences as all-sufficient.

1. In perfect childhood, uncrossed by perverse and chilling influences, and blessed by health, felicity exists — not, perchance, universally, but as a rule. I remember some few pages of my own childhood which were filled with an unbounded felicity, a felicity to be remembered, although it cannot be again realised or explained in

relation to the precise causes that led to it. I have questioned others on the same point; and although the response was much more frequently in the negative than I expected to find it, and although the inquiry has often laid bare a recollection of misery rather than of felicity in childhood, it has yielded, certainly, a majority on the affirmative side. The evidence is sufficient to prove the reality of the phenomenon in at least one stage of life.

2. There are, again, men and women who, by some fortunate heredity of constitution, go through long, trying, and eventful careers with perfection of felicity. Dr. Joseph Priestley was one of these fortunates. 'I was born,' he says, 'of a happy disposition.' And so this man, through a life of struggle and tempest such as few men have known, was ever in felicity. In his child life he loses his mother. He leaves his home, and is domiciled with an aunt, whose gloomy tenets would drive some natures to the deepest melancholy. He passes through severe changes of thought on solemnest subjects. He becomes a preacher, but, owing to a defect of speech, cannot display an eloquence he knows is in him, and, tossed from pulpit to pulpit, penniless, is forced to teach that he may live. He becomes half friend, half librarian of a nobleman, by whom he is petted at first, and then, with the capriciousness of power, is turned off, as a once-favoured dog might be, without a word of explanation. He makes one of the grandest discoveries of the century, and lives to see the discovery accredited to another man, to whom he communicated it in the most open manner. Suspected of sympathising with children of liberty, he is made, under

the instigation of a rival preacher, the victim of a furious mob, which burns his house, and all his precious papers and treasures, wishing him heartily the same fate. Escaping to London, he is obliged to hide from enmity, and, cruelest cut of all, is disowned by and cast out of the learned society whose work he has helped to immortalise. At last, driven in his old age from his native country, he goes, forgiving every one, to a foreign and distant land, to die there in perfect peace.

Such changes as these, such oppressions through every stage of life, would kill a multitude of ordinary men. Yet here was one who went through every phase of suffering with felicity. His friends, one and all, bear witness to this fact from their objective side. He personally testifies to the same, and explains the reason: 'I was born of a happy disposition.'

We gather from such instances as these—rare, it is true, but reliable—that in the range of physical life there is a felicity due to heredity; to some combinations of ancestry, which, being repeated, would lead to the birth of an almost new race amongst which Priestley's own maxim, 'the greatest good for the greatest number,' would be the common blessing. For, that which has once been born may be born again, and by birth become universal as a progress. If one man can hold felicity in his hand all his life and under all adversities, why not all men?

3. There is a third proof of felicity which comes within the knowledge of the majority of mankind although it is not universal, for I have known a few who have afforded no evidence of it. This proof consists

of the sensation felt, I repeat, by most persons of a sense of peace, tranquillity, and, in a word, felicity, which, in consequence of its abruptness and the sharp contrast between what has gone before, is a cause of extreme surprise. In such moments the actual cares of the world, cares heavy and sorrowful, sit lightly; the impossible a short time before becomes the possible or the easy. Dark forebodings which have pressed almost to despair pass away, and the future is roseate with prospect.

There are few now present who have not experienced this curious change towards felicity. They may say that it is a fleeting change, and that may be so; but the fact is certain, and is also immeasurably instructive; for if felicity can be obtained for one day, for one hour, why not for all days, for all hours?

These flashes of felicity are, I have said, sometimes startling from their abruptness. They are at other times equally startling from their intensity, and from the relief they give to the opposite depression, from which they stand out in contrast.

In speaking of this contrast, and of the advent of felicity after extreme depression, the common terms used to express the conditions are singular. The depression is almost invariably described as a physical weight, and felicity as the removal of a weight which, like a physical burthen, has oppressed the body, and, in extreme instances, has bent it low. 'He is bowed down with sorrow' is an expression as true as it is striking. Bunyan seizes on this physical truth. His pilgrim, while yet wanting felicity, is troubled with a burthen which weighs upon his back night and day, felicity coming when that burthen falls from his shoulders. The illusionist has here defined what he himself had felt, and hence the force of a description which every man and woman who has read Bunyan has, with very few exceptions, recognised. Felicity is lightness from burthen. The common folk call it lightness of spirit, light-heartedness, as being lifted up above the common fate of daily oppression and daily sorrow. The terms define the state.

When felicity is most absent, the sense of depression shows itself in other ways, which indicate the physical process, and suggest the ponderable nature of something that tells on the body and on the mind. In worst states of depression the faculty of memory is often overburthened with labour of details, long stored up, which are remembered, re-arranged, and re-conjectured upon with painful and accurate precision. The thoughts undulate, and great waves seem to overwhelm another organisation belonging to the man himself, yet lying afar back, and obscured by these rolling tides, dark, dense, material, weighty. With felicity all these waves of deathly pressure pass away. The memory is charged with no recurring scene of sadness. The calculated difficulties do not appear. The organisation which lies in the shade becomes brilliant, and the future is charged with hope.

These phenomena constitute both by reality and contrast what may be called the full-grown subjective proofs of felicity.

4. Lastly, there are certain objective proofs which lookers-on may observe if they will notice others, and which, as independent evidences, are perhaps the most

reliable. A good perception of character and, if I may say, diagnosis, leads the looker-on to note and know the symptoms of felicity in others, for the symptoms are clear. In the wake of felicity the pulses are regular, tonic, free. The breathing is natural. The eye is bright and clear. The countenance, even in age, is youthful. The appetites are keen but orderly. The judgment is sound and joyous. The muscular bearing is firm, co-ordinate, steady; there is no indication of carrying a load on the back, nor of oppressive sinking exhaustion.

In the above few passages I have sketched out, as far as I dare in the allotted time, the phenomena of the felicity we are now considering. I have entered into no metaphysical subtleties in definition, but have rested on every-day experience; and having thereby, I think, afforded evidence of the fact of felicity, I pass to the thought how to extend this state—a thought which, according to my view, is eminently sanitary and practical.

PHYSICAL CONDITIONS INFLUENCING FELICITY.

To arrive at the true mode of working for this object we cannot do better than survey, in the first place, the conditions under which the phenomena of felicity, and of its opposite, depression or infelicity, are manifested.

Atmospherical Conditions.

By a sort of general impression the weather is believed to exert a peculiar influence for and against the phenomena of felicity. In this view there is undoubted truth. An *increase of the atmospheric pressure* or *a decrease* may be a cause of felicity. In ascending from valleys to moderate heights there is, up to a certain distance, a distinct effect of the kind. So definite is this action that I know of one person who, under some conditions, feels life is a load too hard to bear, but who, in a dry bright mountain region, to which resort is often had, throws off the despair altogether and lives a new life. In the nicely-adjusted balance of atmospheric pressure against animal circulation of blood, the circulation is relieved by a moderate removal of pressure. But if such removal be too great, if the organs of the body become congested from the removal, as they may be, the spell is broken.

The brightness of mind induced by removal of pressure and freer circulation is, however, bound by other conditions. Dryness must accompany lightness of air to produce the state favourable to felicity.

There may, again, be conditions in which a slight excess of pressure may be conducive to felicity. In regions where the land is low, compared with the sea level, a slight atmospheric pressure may be advantageous. The air is usually drier under pressure, the wind bracing, and the vital organs charged with blood—conditions essentially favourable in low-lying districts to the whole of the communities that occupy them.

There are *electrical conditions* of the atmosphere during which felicity contrasts strongly and strangely with the depression incident to other conditions. My friend Mr. Hingeston, of Brighton, has very beautifully connected these varying states of atmosphere from electrical influences, and these varying states of mind with cloudland. He reads in the clouds the outward and visible signs of the mental state. The large white-headed cumuli that collect in clear bright days are rotary storms of hail, rain, or thunder, gyrating from left to right. Several of these gyrating storms keep marching onwards in alternate spaces, marshalled in vast circular array, and rolling round a circumference of bright translucent calm.

On the approach of one of these masses of vapour the mercury of the barometer first falls, and then rises with great rapidity.

The accompanying and residual state of the atmosphere is congenial to health. Now the debilitated experience favourable reaction, and the mind is serene and happy. The air in these moments is antagonistic of disease.

With the breaking-up and dissolving of these large cumuli there is electric action, and most likely explosion, just as the vapour is being condensed into water. The entire atmosphere now changes; everything is dull and grey; the so-called dyspepsia prevails; the acid indigestion of gouty habits; the scrofulous, indolent, and pitiable host of 'never-wells.'

Thus, continues my friend, the sensorial effects of the electrical fluid are proof paramount of its pathological, physiological energy, and the various forms assumed by the vapours condensing or dissolving in the air—clouds — may be considered not only as picturesque beauties in the landscape, but also as criteria for judging of some of the most potent effects resulting from the operation of an experiment, silently and delicately performed upon the functions and sensations of animated beings.

Cold and *heat* each play different parts in production and reduction of felicity. A dry and sharp cold—what is called a bracing cold—exerts a gentle pressure on the surface of the body, which fills the nervous centres with blood, and helps to felicity of mind. A long and piercing easterly chilling cold checks circulation, robs heat, and produces even melancholic sadness. A dry genial warmth acts like a bracing cold; a long warmth with moisture checks the vital action, and produces a degree of depression which may be as intense as that which is induced by prolonged exposure to cold.

The seasons of the year which are attended with least exhaustion of the body are those which favour felicity. When the exhaustion of the winter and depressing spring months has been removed by the warmth of a genial summer and autumn, the time is most favourable for serenity of mind. On the other hand, the exhaustion of winter and spring induces depression, and is no doubt the cause of that melancholy which renders the months of April, May, and June the maximum period of deaths by suicide.

Purity of the atmosphere is an unquestionable aid to felicity. The comparison of children living under differing circumstances is sufficient proof of this fact. Children in an open well-ventilated schoolroom, how different are they from those who are immured in the close overpacked dens which are miscalled schoolrooms! The felicity of the children of the well-to-do who live out-of-doors, and even of the children of the fields and open streets, compared with the felicity of those of the small trader whose back parlour is living-room and playground; the felicity of the man or woman who leads an outdoor life, compared with the felicity of those who live in the close office or workroom, how entirely different!

Foods, Drinks, Narcotics.

There are still other agencies which bring or which check human felicity, and which are as purely physical in character as those above recorded.

There are substances which taken into the body produce strange contrasts in respect to felicity and depression. Foods well cooked, foods carefully selected, foods supplied in sufficient quantity to sustain the body equably in all its parts, but so moderately as never to oppress the nervous digestive powers, conduce to felicity in the most telling manner. As a rule all agents which stimulate—that is to say, relax—the arterial tension, and so allow the blood a freer course through the organs, promote, for a time, felicity, but in the reaction leave depression. The alkaloid in tea, theine, has this effect. It causes a short and slight felicity. It causes, in a large number of persons, a long and severe and even painful sadness. There are many who never know a day of felicity owing to this one destroying cause. In our poorer districts, amongst the poor women of our industrial populations—our spinning, our stocking-weaving women—the misery incident to their lot is often doubled by the use of this one agent.

There is another agent more determinate in its effects and contrasts than tea, and that is wine. I am a total abstainer, but I am, I trust, an honest observer also, and I confirm from direct observation the old saying that 'wine maketh glad the heart of man.' If it did this and no more, I should say, Let the felicity of wine remain to the world. Wine, like the alkaloid in tea, relaxes, lets loose the channels of the blood; gladdens

FELICITY AS A SANITARY RESEARCH 287

like the ascent of the mountain side; gladdens like the gentle atmospheric pressure which forces more blood on to the internal parts. But—and here, alas! is the rub; carried a little beyond the right mark, the felicity from wine passes into folly, the folly into feebleness, the feebleness into stupor, and the stupor into a depression the reaction from which is the bitterest, the most persistent.

Tobacco is another of the substances used to produce abeyance of anxiety. Tobacco is said to soothe irritability without stimulation; but it leaves in many persons long depression, coupled, generally, with an appetite for a renewed indulgence in it, which becomes intense. The confirmed smoker, who can stand out against indirect effects, whose taste for food and whose digestive endurance are little injured, is kept during the whole time he indulges, in the state of suspension. He does not enjoy felicity, but for the time experiences a relief from infelicity. My own experience, on the whole, is opposed to the indulgence, for I tasted it for a long period of my life, and have long observed the effect of it on others. To the aged it gives, I confess, a negative existence, which, when the mind is not filled with choice or refined or cultivated pleasures, makes time less wearisome. To the man who engages in work of great excitement and of a mental kind it brings a joyless repose. But, on the whole, it is a bad, and sometimes a fatally bad, indulgence. I have once known a man die directly from its effects; and how many I have seen injured I cannot say, but a large number. Again, I have seen many much depressed by it; so that I dare not put it forward, at its best, as a promoter of felicity. The world, I must conclude, would be happier if tobacco were unknown or unemployed.

The habitual use of opium for the obtainment of felicity is of the same erroneous character. The opium-smoker, the opium-eater, tells us of certain dreams and phantasies which are, for a moment, felicitous wanderings of the mind. I have visited the opium-dens to see the effects; and whatever the dream may be, subjectively, it presents to the observer no sign of felicity. The expression of the opium-smoker is one of restless and intense anxiety. He looks like a man in a dream of misery. His eyes are joyless, his features contorted, his skin colourless or dark, his pulse slow and labouring, his breathing hard and heavy; and when from the half-struggling consciousness he wakes to reason, the dream he describes is too confused to be accepted as a dream of felicity. Then he falls into dejection, which deepens and deepens, until the desire to return to the cause of the dejection is too overpowering to be resisted. To opium-eating and to the subcutaneous injection of morphia the same description, with some modifications, on which I need not dwell, is perfectly applicable. From the use of such an agent as opium there can be no result of human felicity. There could soon be produced by an extension of the use a madder world than now exists and a more miserable; a happier, never!

And this saying, according to my knowledge, extends to all narcotic substances. There are some, like methylic ether and nitrous oxide gas, which produce for the moment infinitely more refined felicity than those I have specifically named; but in the end the results are the same. They all create a craving for themselves in those who become habituated to their use, and engender a new and unnatural constitution.

Constitutional Influences.

I have dwelt thus far on influences of a purely physical kind in their relation to felicity. I have put these influences in contrast as affecting the state of felicity, and I have touched on some agencies which are used to produce a mock felicity. I must move from these to a brief consideration of certain influences of a different nature which affect us for or against our happy or felicitous existence.

Temperaments and Heredities.

There are some constitutional differences determined by temperaments which are of first importance. Of the four primary temperaments—the sanguine, the nervous, the bilious, the lymphatic—and of their relation to felicity, a volume might be written; and I have collected the facts relating to the temperaments of over a thousand persons towards such a work. I must not here touch, however, on any detail. I must be content to record as a general fact that the sanguine is altogether the happiest temperament, but not always the most sustained; that the dark or bilious is the least happy in early life, but is often, in later life, more serene; that the nervous is a varying condition full of ups and downs; that the lymphatic is, by a negative effect, the most even; that amongst the twenty-four combinations of temperaments the sanguine lymphatic is the most felicitous in respect to physical pleasures, and the bilious sanguine and the bilious lymphatic in respect to intellectual; that the nervous sanguine is the most irritable, and the nervous lymphatic the most helpless and miserable.

The moral influences and impressions affecting these natures are, from first to last, potent for good or for evil. In childhood the future history of the felicitous or infelicitous after-life is usually written. A few, born, like Dr. Priestley, of a happy disposition, fight through all adversity, filled with a magic soul of felicity; but they are very few and are commonly, though I dare not, in the face of natural truth, say always, good. They, even in childhood, are not affected as others are. In the schoolroom or playground they are comparatively happy.

Felicity as a Physiological Quality.

As a rule, the tendency to felicity or its opposite is planted in childhood. The parent, the schoolmaster, the schoolmistress, hold the book not of learning simply, but of fate. To the imitative mind of the young, absorbing all that the senses can carry to it and the nervous centres can retain, the character of the presiding mind, ever present with joy or sorrow, justice or injustice, love or hate, cruelty or mercy as its qualities, is the beginning of the end.

Let me, as bearing on these matters of thought, not diverge from, but converge to, our present study, by a reference to felicity as a physiological quality.

Of the two living natures with which man is endowed, and which distinguish him from the lower creation, —the pure animal and the pure intellectual natures,— felicity belongs to the animal nature. An intellectuality that would separate man from the animal would leave him beyond either felicity or infelicity.

Felicity, in fact, is not an intellectual quality: it is not centred in the brain. It is not a quality which a man can think himself into, or reason himself into, or directly will himself into. It is, like the beating of his heart and the circulation of his blood, a vital process going on independently of his volition. He can by rude process destroy it for himself as he can for others. He can kill it for himself as he can stop the motion of his blood by stabbing himself to the heart; but still the quality is so independent of himself that he is often forced to be in felicity by things and acts and circumstances which his reason scorns. 'Why does this fool of a book make me happy?' said a hard and suffering and miserable patient once to me as he pitched his 'Pickwick' to the foot of the bed. 'Yet it is the only thing that does, while all the time I know that such a set of asses as these Pickwickians could never possibly have existed.'

By experience of what seems to increase felicity, to lighten misery, to make misery, we often confound felicity and sorrow with intellect. This is merely our own external looking upon external manifestations of internal phenomena which we know we cannot influence in the same way as we can teach a lesson or convey a fact, but which we strive to control because we think we must do something, even for the uncontrollable.

Felicity and infelicity are not intellectual faculties; neither are they passions; neither have they any direct relationship to physical pain. They are the only true emotions. The man who is destitute altogether of felicity is not, of necessity, intellectual, passionate, destitute of passion, or more or less sensitive to pain than any one else. The most intellectual may be the most miserable;

the most silly and inconsequent may be the most blest with felicity. The worst instance of extreme, I may truly say harrowing, misery I ever knew was in one whose clearness and calmness of judgment was a subject of general admiration, but who had never, he told me, known in all his life an hour of felicity. The man most replete with felicity I ever knew was one endowed with no intellectual supremacy at all, and who was all through a long life a veritable child.

The centre of the emotion of felicity is not in the brain. The centre is in the vital nervous system, in the great ganglia of the sympathetic; lying not in the cerebrospinal cavities, but in the cavities of the body itself, near the stomach and on the heart. We know where the glow which indicates felicity is felt; our poets have ever described it with perfect truthfulness as in the breast. It comes as a fire kindling there. No living being ever felt happy in the head. Everybody who has felt felicity has felt it as from within the body. We know, again, where the depression of misery is located; our physicians of all time have defined that, and have named the disease of misery from its local seat. The man who is always miserable is a 'hypochondriac'; his affection is seated under the lower ribs. No man ever felt misery in the head. Every man who has felt misery knows that it springs from the body, speaks of it as an exhaustion, a sinking there. He is broken-hearted; he is failing at the centre of life; he is bent down because of the central failure, and his own shoulders, too heavy to be borne, feel as if oppressed by an added weight or burthen, under which he bends as though all the cares of the world were upon him to bear him down.

HABITUAL, SENSATIONAL, AND MORAL INFLUENCES.

There are numerous influences which in an indirect way tell upon felicity, for it or against it, as distinctly as those grosser agencies of which I spoke in the earlier part of this section of my address. To these I would now invite attention.

The influences to which I am about to refer are, in reality, purely physical in their action; although they are commonly known as habitual, sensational, moral, or mental influences.

I notice, in the first place, that felicity is always favoured by sufficiency of *rest and sleep*. Bad sleepers know no felicity; but they who in childhood and old age sleep ten hours, in adolescence nine, and in full age eight hours out of the twenty-four, and that soundly, are mostly well favoured with felicity. They may be exposed to causes which are opposed to felicity, but even then the causes are feebler in action than they otherwise would be. I put sleep in the first place as an aid to felicity because it comes first. I have no knowledge of any instance in which a person who slept well was altogether devoid of felicity. The beneficent action of sleep is, however, indirect. It is due to the physical and mental strength which it confers on its favoured child.

Strength of body secures felicity. Persons comparatively weak of mind may, with a good physique, be happy; but very few who are weak of body have any long tastes of felicity. We may take it all round that the feeble of all ages are unhappy. It is a matter of common observation that persons who are so unfortunate as to be born deformed of body, though the defect be concealed or hidden, are not blessed with felicity. It need not be the deformity that causes the infelicity, for the deformity may be concealed; the bad health is the rooted cause. Let the defect be from accident happening to a body born of good stamina, and felicity may be the same as in others, despite the acquired defect.

Any sign of inherited weakness is an equal sign of lessened felicity, though it be marked by no physical defect. It has been long observed by physicians that persons who from early life show very large and prominent veins, and thereby a languid circulation of the blood, are never happy, while those of well-knit body are. The observation is true as steel. We physicians know that a sluggish circulation is incompatible with felicity, and that they who show this indication—by such local diseases as hæmorrhoids, for example—are amongst the most depressed of those who consult us. We say of them that they suffer from arrested circulation through the liver, and, doubtless, such arrest is depressing; but the arrest means physical exhaustion, and physical exhaustion is the root of the evil. When the circulation is sluggish the liver is sluggish, the brain is sluggish, and the nervous centres are depressed. In a sentence, whatever prevents physical exhaustion and sustains physical strength sustains felicity; whatever exhausts sustains infelicity.

When the sun of life is high
   All is bright.
When the sun of life is low
   All is night.
Thus we laugh and thus we sigh,
Light and shade where'er we go.

Physical work, when it is carried short of exhaustion, keeps up felicity; and sloth destroys it. But the physical work that exhausts kills felicity. The argument extends to *mental work*. Moderate wholesome mental work is the best of all aids to felicity next to sleep : it strengthens the mind; it softens grief; it lessens care. Carried to excess, it is pernicious and destroys all felicity. Cowper, the poet, was wont to say that no labour is so wearying as composition, and few men possibly felt more unhappiness as the result of mental exhaustion than he. But his was the fate of all who force the brain to daily or nightly repeated weariness.

The influences derivable from sleep and bodily power are purely physical influences; but there are others, called *sensational*, which, through the physical power, have a potent effect for or against felicity. The sense of hearing has the most intimate connection with the vital or animal nervous system. The auditory nerve, as Dr. Bucke has shown, possesses many of the characteristics of an organic or sympathetic nerve; and we all know how many external vibrations which reach the ear affect the digestive system, producing sense of warmth in the body, appetite for food, and feeling of felicity; or, on the other hand, setting the teeth on edge, destroying appetite, and giving rise to gloom. Thus things told affect quickly, often permanently, for good or bad. The sense of sight influences also, but less intensely, and after a different manner. Sights gladden or dazzle, or pall or sicken.

Not to extend an argument which admits of any extension, the senses, as doors and windows through which influences vibrate into the animal organisation, can scarcely be touched by external phenomena without conveying some influence that shall make, or disturb, or prevent felicity. When they convey beauty in sound or picture; when they convey variety; when they convey cheerfulness of act, and manner, and voice, and feeling; when they convey to the soul the idea or story of generosity, of courage, of purity of life and character, of prosperity: then they convey felicity, which, passing through the brain on its way, finds its seat in the vital centres on and near the heart.

With felicity as a sanitary research for my theme, I have striven so far to indicate what may be called the physiological bearings of the subject. I have endeavoured to show that felicity is something that is of hereditary quality; that it is something made or not made by external agencies over which we have little control; that it is something made or not made by many agencies which we have directly under our control.

In these respects felicity stands precisely in the same position as health; in the widest sense it means health, is another word for health. Health is born, and is made and unmade by external agencies which as yet are out of human control. Health is made and unmade by numerous influences which are under human control. Felicity similarly influenced, depends on the good working of the animal or organic systems of life.

I could enter here into one of the most absorbing questions relating to the connections which exist between the lower and the higher human faculties. I could indicate how the lower and higher nervous centres, charged during life with a subtle ethereal medium, communicate with each other and with the outer universe; and how, by the states of this refined intercommunicating bond or sphere, both health and felicity are moved by external pressures, by external vibrations, by external agents taken into the living organism, and by products generated within the organism and diffused through its own atmosphere. But I leave this inviting subject for a more immediately practical application of the few minutes which remain for discourse.

PRACTICAL DESIGN FOR FUTURE WORK.

Over those atmospherical causes which have been noticed as influencing felicity we can exercise as yet no direct action. At the same time, just as we can now divert the lightning from its fatal course, we can indirectly effect good. We can prevent, as far as our teaching is effective, the erection of human habitations in dank and humid places, below the sea-level, or in dry and arid spots. We can protest, and if we are clear and reasonable in our demands we can successfully protest, against the construction of new towns on melancholic foundations, and can explain the choicest places for felicitous existence in so far as foundation is concerned.

A report from such an institution as ours, or from any learned sanitarian, on this one subject, addressed to those children of enterprise who are colonising the worlds that are to come, would affect, if it took root at all, all the generations of the men to come in those new regions, and to a large extent the felicity of the whole future human family. In Africa at this moment the seed of new life that is being sown will largely be seed of felicity or seed of sorrow, according to the selection of the sites on which the new and great centres of life are constructed.

Nay, in this direction man himself may divert the operations of nature herself. He may change her surfaces as he cuts down her forests, or plants new forests, or alters the courses of rivers, or makes new courses, or fills up valleys, or lowers or raises mountains, or connects or disconnects oceans. With the mastery of the surface of the planet in his power, man may, in fact, make what regions he pleases for good or for evil.

The earth is the freehold of man.

If the natural air which man makes not and invented not may be to a large extent utilised for felicity, how much more easy is it for him to remove the unnatural which he himself makes, so that instant advantage of what is provided for good may be rendered serviceable!

Here our voices should be heard in a tone not to be mistaken. We shut up our young in closest rooms of close towns; we shut up our men and women by the millions in close shops and factories. Some one million of us in these islands who call ourselves, with ignorant irony, the ruling classes, shut up some twenty-five millions of the people, including wives and children, in walled-up atmospheres, where atmospheric purity is unknown; where cold and heat oppress, where food is whatever can be got; where drink is what can supply a false delight for a certain sorrow; where marriages re-establish misery; where good sleep is impossible, where physical strength is so impaired that a perfect body is not to be found; where exhaustion from work is the daily cross; where things and objects of beauty are rare as angels' visits; where in the selfish race to barely live generosity is impossible; where in compressed homes purity of mind is a thing the purest can scarcely maintain; where variety is replaced by the dead monotony of unchanging sounds for the ear to hear and of scenes for the eye to see; where fear dominates over courage; where hope has no chance; where prosperity is so little known that the worn-out life has no expectation this side of the grave; and, where death is so busy that three die to one of the more favoured communities. We, one million, I repeat, shut up our twenty-five millions under these conditions, and wonder why those millions know nothing of felicity; why they are peevish, reckless, melancholy, sometimes drunken, sometimes rebellious and ready to run after any leader who shall promise to guide them into a happier sphere, however little removed from that in which they are. Wonder! The wonder is how human nature can bear such a famine of felicity and live as if it only lived to die.

To give the boon of felicity to these masses, we, not as revolutionaries, but as laborious workers for them and the thing most wanted, have, I trust, come into existence, and not in vain.

It is better for this work that we should be as we are —men of science rather than men of politics. We are, then, in the first and true place. We are educating politics as well as men by what we teach—an education which, in the present dense state of political darkness, is the noblest work we can possibly execute, until those who rule understand common human nature.

In the direction of education the sanitarian teacher should, I think, begin to study this psychological side of humanity: what human nature can and cannot bear; how much pressure can safely be put on humanity without danger of explosion; how much felicity can be secured by removing that danger.

Beyond the task of inculcating what are the necessities calling for pure atmospheres within and without the body, we can, most appropriately, explain what agents taken into the body are for felicity. We can teach what temperance in all things effects in this direction. We can use our earnest will in declaring for abstinence where nothing but abstinence is the safe line of conduct. We can denounce every indulgence that undermines and mocks and destroys the blessing. We can insist on cleanliness—cleanliness of body, cleanliness of mind. We can show how flowers and plants grow for health as well as beauty, and why for both health and beauty they must join the home circle if felicity is to enter it.

These are all sanitary questions, and they all, as one, bear on felicity.

We may educate again in another direction. It has been shown that some exceptional men are born of a happy disposition; and it might have been shown, on the clearest evidence, that multitudes are born of an unhappy, nay, miserable, disposition. We could easily by our researches describe what are the lines of heredity for the happy and for the unhappy dispositions. We could, with this discovery in our hand, with certainty of being listened to and attended to, impress on the people the truth that marriages ought neither to be matters of chance, nor matters of mere monetary convenience, nor, indeed, matters of mere insane so-called love; but that the marriage tie, extending its influence into the future, and being no bond and seal of diseased heredities, should be the bond and seal of a healthier and happier racial progress in every succeeding generation. Now that our women are, by good fortune as well as good policy, made legally masters of their own property, this sanitary question involved in marriage was never so likely to be one of scientific value as it is at this moment, and as it will be in coming days.

We can teach forcibly and faithfully on a different topic regarding which we have a large amount of information collected. We can adopt and urge with all our power our veteran Chadwick's advice to those who are wanting to instruct the young, that it is the perfection of prudence first to live, then to learn. We can insist that, inasmuch as felicity is impossible under mental strain, it is fatal work to press on the young mind the excessive labour which is now in all departments making cram, cram, cram the footing for knowledge. We can also tell the adult man struggling for the bubble reputation that broken sleep, and disturbed brain, and wearied muscle, and labouring heart can never exist with felicity: that *sanitas* and *vanitas*, separated by one letter only, are as the poles, apart from each other; and that *Sanitas sanitatis, omnia sanitas* will never be established until *Vanitas vanitatis, omnia vanitas* is blotted out.

We can instil yet one other lesson, last but not the least, into those foolishest of the foolish of the world who think that riches and repose and power are synonymous with felicity. We who have read the human heart, and have learned, by closest observation of fact, the lives of men, can tell these that no success, no wealth, no power, gained by torture of effort for it, ever brings felicity, and that the Horatian verse :—

> He who would hold the golden mean
> And live contentedly, between
> The poor man and the great,
> Ne'er feels the wants that pinch the poor,
> Nor plagues that haunt the rich man's door,
> Embittering all his state;

is as true in these days as it was in the days of the philosopher who wrote it.

Felicity as a sanitary research.—With all respectful thought I leave it on your memories. You may perchance think of the idea as an enthusiasm, a vision. Never mind, so long as you think of it. It will grow upon you as a study and flow from you as a project if it once take root. It will strike you, in time, as the *summum bonum* of sanitary labour—a re-echo of the Divine declaration, ' On earth peace, and good will towards men.'

CYCLING AS A HEALTH PURSUIT—
PHYSICAL AND MENTAL.

To the human family the art of cycling is the bestowal of a new faculty. I am not an accomplished cyclist, yet I find that by means of the simple machine, the tricycle, facility of progression by my own muscular powers is fairly doubled, while half the weariness incident to progression on foot is saved. If I walk ten miles in three hours—a fair pace—I am tired; my ankles feel weak, my feet sore, my muscles weary; so that after the effort I am unfitted for any mental work until recruited by a long rest. If I go the same distance on the tricycle on the same kind of road, I find that an hour and a half is the fullest time required for that journey, and when the task is finished, instead of feeling a sense of fatigue, instead of being ankle-wearied and footsore, I am agreeably refreshed by the exercise, and ready for study or other mental occupation.

In the earlier periods of my professional career riding on horseback was a necessary part of the daily life. I had learned before then to ride without a saddle and to become habituated to all paces; and, until very lately, I have kept up horse exercise from the love of it; so that still, with a little renewed training, I can take a ride on horseback of five-and-twenty miles without excessive fatigue. But I would much rather be forced to ride forty miles on a tricycle, if the riding were a matter of choice, and the question of fatigue the point that determined the choice. This personal knowledge is tendered because it is the most practical, in support of the statement that the art of cycling confers on those who learn it a new faculty of locomotion. It is also intended to be a moderate proof in point; for, if I were to follow it up by the record of what accomplished cyclists have achieved, it might be proved that two or more faculties had been added, and that when a man can wheel himself at the rate of twenty miles an hour, and a woman can wheel herself a hundred miles a day, there is found a new machinery in the human body itself, a new set of muscles almost, a new skeleton leverage, a new kind of volition.

We are entering on a new era in locomotion. To those of us who have studied the question of muscular motion physiologically, it is quite clear that concentration of power in great engines is not the ultimate, as it is not the natural, design for progression, because it is not primary. It seems a wonderful thing and an easy thing for hundreds of persons to be moved by one steam-engine. It seems like starting and moving from a beginning if we shut our eyes to everything until we see the engine and train before us. But let us go back. Let us think of the enormous amount of muscular power that has been employed to extract the materials out of which that engine is made, to construct the engine, and to dig out from the earth the coal that feeds the furnace; and all by living power. Then the questions spring up, how much human labour is actually saved, how much of the severest labour is added to mankind by the engine? And soon another question springs up—namely, how long will men consent to be engines for engines? It is hardly in human nature to suppose that men will long continue to hold such a position. Will they hold to it in these islands for another fifty years? I very much misunderstand our rising countrymen if they will. Sir Humphrey Davy, in the later part of his life, said that long before coal was used up men would know how to burn water. That is probably true; but what about coal-working? Long before coal ceases the coal-worker will cease, perchance before men learn to burn water.

We want, therefore, to return quickly to first principles: every one his own locomotor against time.

By the simple machines, bicycles and tricycles, we are returning to first principles. We are endowing every person who can use these machines with a new and independent gift of progression; and to what extent this art will proceed in a quarter of a century, if it makes the same progress that it has made in the past twentieth part of a century, he were indeed a bold man who should venture to predict. I have already said, what I now take occasion to repeat, that the art of flight will be the practical outcome of the grand experiment which is now going on; for, when a machine can be reduced in weight to thirty-six pounds, and when such a machine can be propelled on a good track twenty miles within the hour by human limbs, carrying the man who propels it, there are not many removes to the capacity of driving-wings or air-screws at a sufficient rate to afford support to a machine on the air. I think many persons will, indeed, live to see a partial development at least of this kind.

They will, I mean, see constructed a machine which will be partly sustained by the air and partly by surface of water—sea, or lake, or river—and which will skim over such surface with just sufficient friction for steerage power and no more; in short, a flying canoe or boat which, elegant and useful, will at one moment, like a nautilus, run with the wind, and at another skim the water, independently of wind, like a sea-bird.

As we stand at present, we have, then, to recognise firstly, an accomplished fact, namely, that men and women can do such remarkable feats in progression, that they are now the swiftest and most enduring of all land animals; and secondly, a possible advancement, namely, that men and women may soon rival in locomotion animals which speed on their way through the air. These thoughts suggest that the time has come when an inquiry is demanded on the questions whether we are now employing what has been done to the best of purposes, and whether the future outlook is in every sense satisfactory. It appears to me, as one of those who take part in the development of this great movement, that the best is not being done, and that a new departure is immediately demanded in order to keep the progress of the art of cycling in a proper position in respect to its advancement and its usefulness.

If you ask a cyclist why he takes part in the movement, and what his interest in it means, his answer in nine cases out of ten is conveyed in the word 'sport.' So cycling is called by its advocates 'the sport'; so the ambition of cyclists is to appear in the sporting columns of the newspapers as winners in the different competitions; so the men, and for that matter the women too, who have made what is called the best 'record' are thought to be the choicest representatives of the cycling community.

I do not write these lines to complain against what has occurred in this way up to the present time. I am perfectly aware that the restless activity of those who have made cycling a sport has stirred up manufacturers to the exercise of their finest skill and choicest work in the matter of construction, and that the registry of the 'record' has produced not only the registry of the best riders but of the best machines. This is all right in its way, and it may fairly be urged by the friends of the sport that but for it none of the grand mechanical successes would have been brought forth. In like manner the gentlemen of the turf argue that horse-racing keeps up the breed of the horses and the skill of the riders.

It is fair to go still further in concession to the modern cycling fraternity. It may and ought to be admitted that the racing they have encouraged has had in it no gambling characteristics; that the rewards for winning have been of the simplest kind; that money-racing has been generally discouraged, and that men who would make a living by competition, and who come, in consequence, under the name of 'professionals,' are excluded from clubs and from competitions in which the amateur, as distinct from the professional, sentiment prevails.

These acknowledgments are as frankly as they are necessarily made. I would not for a moment wish to interfere with that competitive sport which, in a natural and healthy form, would keep healthy improvement at all times in its place; but I and many others think that, if the art of cycling is to run altogether into 'sport' and into a matter of comparative excellence of speed and endurance—in other words, into pace and pluck—it will of necessity lead to results which will ruin it in regard to its health-giving quality, its tone, and its usefulness.

Let me deal with the first of these points as the chief one, and as governing the rest.

The greatest benefit hitherto that has sprung from the art of cycling has been the good it has effected on the health of those who have practised the art. I really know of nothing that has been so good for health. Men and women who, ten or fifteen years ago, were immured from one year's end to another in close towns, and who had little experience of country air and country landscapes, are now seen rushing by thousands out of the towns into the country, and enjoying all the natural advantages resultant from so important a change. By this course they have also made their own houses in the towns healthier, because they have left the confined intra-mural spaces in which they existed to be ventilated and recharged with fresher, if not with fresh, air. They have learned how to ventilate their own bodies, and to imbibe an air free of injurious vapours and particles, while they have developed a freedom and a strength of limb, and a mental pleasure and escape from care, which have been useful alike to mind and body. In addition, they have been gainers of many good and serviceable mental qualities. The little bit of risk or danger which attends the expedition has called up courage, attention, decision, and presence of mind; while the desire to perform some predetermined task has taught them how to keep up endurance. The true cockney has been quite transformed by the art of cycling, and in a very few years will be unknown even in Cockaigne.

We must accept this growing change, however, with all its drawbacks, and with no drawback more conspicuous than the fever and strain of competitive struggle. It seems to me that young and old, male and female, weak and strong, are all going wrong on this mania about records. If I could publish the letters that are sent to me making inquiries on the question, and what can and what cannot be undertaken by particular persons under particular circumstances, of what persons are wishing to attempt, and of what persons have attempted to do or at all risks have done, the reader would not wonder that I am getting a little anxious about the future of what might be one of the most valuable of all the physical exercises ever invented.

The following are fair specimens of the dangers in view :—

A gentleman seventy-eight years of age has started a tricycle. He finds, to his intense surprise, that he can ride from Brighton to Lewes without fatigue. That is about eleven miles. In a few days he discovers he can go there and back in the day without fatigue. A few days later he tries to do the same distance against time. He can do it in four hours. But there is a young fellow he knows—who, by the way, is only sixty years younger—who can do it in a little over two hours ; and, why should he not come near to that mark also? It is a mere matter of practice and skill. So he does his best; and having little elastic tissue left in him fitted to give his lungs and blood-vessels due elasticity, he finds himself jarring all over like a ramshackle old bone-shaking bicycle. In fact he cannot get over the 'jolting,' and dates a good deal of mischief from the jolting, as if the machine, and not his ancient human machine, were the thing at fault.

Here let me, in parenthesis, put in a word before I forget it. Whenever a middle-aged rider, mounted on a good machine and riding on a good or fair road, feels jolting, it is the rider that is jolting, and he has ridden enough for that time. If he continues to oppose this admonition, he will be shaky and uncertain in movements and resolution for many after hours or days.

Another gentleman, not quite so old as the last-named, makes up his mind against distance. It is his firm determination to ride from London to Bath, one hundred miles, in a day. It is merely a matter of starting in good time in the morning, and taking the journey in easy stages. He does thirty-three miles as a trial right off, and feels 'as fresh as a daisy.' After a little rest he is surprised to find he has a curious sinking at his stomach and can take no food. He says he was stopped in his effort by a bad fit of indigestion, which spoiled his riding for a fortnight, and rather set him against it altogether, because he could not go from London to Bath in a day like other people.

This suggests another mem. of much practical importance. Whenever a rider feels a sense of stomach-exhaustion from riding, is unable to take food with appetite after a ride, or digests his food slowly after a ride, he is told as plainly as these words can tell him that his nervous tone at the centre of life, the stomach, is exhausted, and that he had better do no more until he has rested completely and restudied the fable of 'The Belly and the Members.'

A middle-aged gentleman, who is engaged in sedentary occupation, and who is subject to occasional attacks of rheumatic gout, is advised, very properly, to take to the tricycle. He takes to it, and, learning to do eight to ten miles per day, is quite astonished at the result. He feels like a new man. His spirits are so light, he walks so well, he sleeps so well! He could not have believed that such a change for the better could have been effected. He waits for a holiday that he may get over a good deal more ground. Previous to the holiday, and indeed preparatory to it, he keeps close to office-work, and then, getting clear, he starts off for a tour. The first day he gets over his thirty to forty miles, perspiring very freely. The next day he tries to repeat the experiment, and then, strangely enough, as he thinks, he is visited with a smart touch of his old enemy the rheumatic gout, and is obliged to give up.

A third mem. of much practical meaning is suggested by the above-named experience. Whenever a man of sedentary habits who is greatly benefited by the tricycle exercise, taken in moderate and regular efforts, finds that he is getting premonitions of rheumatic affection—pains in his wrists and in other joints, feverishness, acidity, and restless desire for rest while cycling—he has an intimation of having done too much. He has produced a degree of waste of his own muscular fibre, beyond that which he can freely eliminate, and has made himself rheumatic by his experiment.

These admonitions, which are drawn from direct observation of natural occurrences, in respect to persons who are of or beyond middle age, extend in another direction to the younger members of the cycling fraternity.

Cycling, as an exercise and a sport, has not been in fashion long enough to enable us to see what may be the effects of it on the young who practise it excessively, and under special circumstances of strain and fatigue. What has been made out so far is beyond any expectation in its favour. No other mode of progression at the same rate and for the same distance has ever been accomplished with the like freedom from actual exhaustion. To make a hundred miles a day on ordinary roads on a bicycle is now considered commonplace amongst practised riders, some of whom indeed smile at two hundred as nothing very particular. One hundred miles in the twenty-four hours on the tricycle was once looked upon as remarkable; but in time Mr. Marriott rode one hundred and eighty-three miles, and he has since beaten his own record by a ride of two hundred and nineteen. Even ladies have cast in their skill for these sharp and long tricycle rides; and one of them, Mrs. Allen, has accomplished one hundred and fifty-three miles within twenty-four hours.

These results of a single day's work take us fairly by a surprise, which is only overmatched by what has been effected in short runs against time.

In short runs against time on the track, over twenty miles an hour have been made on the bicycle, over sixteen miles on the tricycle. In short runs against time on a level road, eighteen miles an hour have been made on the bicycle, sixteen on the tricycle. I have been a witness of the fact of twenty-five miles being accomplished on the bicycle on a hilly road, within one hour and three-quarters, namely, at a rate of fourteen miles an hour, by a rider who had not yet reached his twenty-first year.

The above two results admit of being compared with efforts made in endurance through long journeys, over the track or prepared firm level surface, and over common roads extending from one part of the country to the other, and of the most varied kinds of surface. In some of these efforts on tricycles distances of seventy miles a day have been kept up for thirteen days, and John o' Groat's to Land's End has been traversed at this rate. On the bicycle the same journey has been accomplished at the rate of over a hundred miles per day.

It took the famous anatomist John Hunter fourteen days to ride from Edinburgh to London on a fairly good horse. A modern bicylist in good training would do the same journey in four days. It was considered a wonderful feat that Mrs. Siddons should one night play in London and the next night appear, thanks to the post-chaise, on the stage of the theatre at Bath. In these days a trained lady tricylist might perform the same feat by her own unaided muscular efforts.

It is only fair to give due and proper praise to labours which have made these changes possible. I am quite aware, from having watched the course of improvements throughout, that the success has been attained by the courage and what I may call the sport industry of those who have ridden the machine. I am quite aware that these enthusiasts have been obliged to carry out their work under extreme disadvantages. I well remember reading a calculation from a man of science, who was thought to be by no means a contemptible reasoner, that it was positively impossible for any person to propel himself on the best road at a greater pace and for a longer period than was attainable by the simple act of walking.

I also remember what a machine it was that the early bicyclist, and still more the tricyclist, had to begin upon in the early days of the art. I have seen how all these difficulties have been met one by one; and how the riders of machines, by their contests against time, distance, and endurance, have made the mechanical geniuses keep up with them. I thank those who have effected so much in the way of progress as sincerely as any one of their most ardent admirers.

There is, however, a danger of enthusiasm turned in one direction alone; and this enthusiasm is, I am sure, carrying the young cycling fraternity too fast and too furiously in what they call their sport. It is true that in cycling there is an immense saving in vital organs, as compared with the strain which is put on those organs by other exercises, such as walking, running, climbing, and rowing. In cycling the whole weight of the trunk of the body is taken off the lower limbs, while the concussion produced by the descent of the foot on the ground is saved—a great saving. In cycling there is no strain on the muscles of respiration in any ordinary effort, and therefore the lower limbs do the greater part of the work, and that without strain of tendon or friction on the sole of the foot or weariness felt at the ankle and knee.

Were there not these savings there could be no accomplishment of a tithe of what is commonly performed even by common riders. At the same time it is not all safe. Putting altogether aside the dangers which are apt to occur from falls and other physical accidents, there is in extreme competitive strife a strain which is, I am quite sure, most injurious to the organism, and which will, I confidently predict, tell seriously on the lives of several cyclists who are now carrying their exertions to an extreme length in respect to strength, pace, and length of effort. In one man I can see the clear evidence of premature age, induced solely by over-taxation and persistent training. In another I observe the excessive muscular growth in the muscles of the leg, like that which has been observed in the opera-dancer—a sign which, local though it may be, is a bad sign, as indicating an unequal development of the body, and what we doctors term hypertrophy of muscle, which will in the end lead to loss of balance of power in the affected parts. In a third I notice a pallor and vibration which indicates a disturbance that ought not to exist between the vascular and nervous systems. In a fourth I detect a restlessness and feverish anxiety which bodes no future good. And in many we are beginning to recognise a too nervous interest to everything that pertains to the sport to mean success to the maker of it or to the sport itself.

Most fortunately it has been discovered by the competitors themselves that a perfectly temperate habit, in which the use of alcoholic poisons is excluded, is necessary for the best competition; and in this discovery there lies a vein of safety which is largely assuring. Hard-drinking cyclists would go to the hospital, the asylum, and the grave as fast as their machines could be made to carry them.

Cycling for Mental Health.—A Project.

INFLUENCED by the facts and observations recorded above, I endeavoured three years ago, through the pages of 'Longman's Magazine,' to indicate that there are other contests and other conquests to be won by the cyclist than those which are intended to develop his physical strength, skill, and endurance. There were, I ventured to intimate, some mental contests connected with the art which might call forth powers and abilities of a different order.

The suggestion thus offered was, that the ladies and gentlemen who had time to make cycling a pastime and a healthy exercise should form themselves into an association, society, or institute, for collecting various kinds of information while they were carrying on their excursions in this and in other countries. The head-quarters of such an association should be in London; but there should be local organisations or branches in all the country towns, and a local secretary or honorary secretary in or near each country town, who should be so well versed in local knowledge that he could give every information that might be desired by those who were passing into or through his district. I thought also that such a society should issue a series of short historical or scientific abstracts conveying to inquirers the leading facts that might be wanted.

The parent society or association here suggested ought, I conceived, to be divided into four sections, which should be designated as follows:—

1. The Archæological.
2. The Geographical.
3. The Natural Historical.
4. The Mechanical and Constructive.

In the society as a whole there should be a president, who should hold office for one year, subject to re-election for two or three years, but not for a longer term consecutively; a general secretary; a treasurer; and a council of, say, twenty-four, who should conduct the business of the society between the general meetings at which they were elected.

To the branches of the society there should be the same arrangement—a local president, secretary, treasurer, and council, who should conduct, subject to the central authority on matters of principle, the local organisations.

In the metropolis the society should have its own house or institution, which should be fitted up with all conveniences as a centre to which every member, metropolitan and local, could repair for reference and guidance. There should be in such an institution a good and carefully selected library relating particularly to the subjects that are included in the objects of the association.

This central house might in course of time become the grand centre of the cycling community from all parts of the world. It might be a public treasury of information which would have few rivals, since it would soon receive additions of information and collections of curious treasures illustrative both of the immediate present and of the past and remote past. In the central society as well as in the local there should be held each month, excepting during holiday months, an open meeting of members of the four different sections, which meetings should be presided over either by the general or local president, or by a vice-president, who was eminent for his archæological, geographical, natural-historical, or mechanical learning. At such meetings papers should be read by the fellows after the manner pursued at the different learned societies, and discussions should follow the readings without voting, unless there were any special occasion when the record of the vote was considered absolutely necessary.

The papers read at each successive meeting should, I suggested, follow the order of division of sections. At one meeting the subject should be antiquarian or archæological or historical. At another meeting geographical topics should be introduced on the physical states and peculiarities of places in different parts of the world that have been visited by the reader. At a third meeting the subject should be natural-historical, introducing various observations in natural history, relating to geology, botany, zoology, meteorology,—including information on weather and climate,—and most importantly, anthropology in its widest sense,—man in respect to his different characteristics, modes of life, learning, physical culture, health, in various parts of the globe. At a fourth meeting the matters read and discussed should have reference to mechanics, and especially to those mechanical inventions which are connected directly with the art of cycling. At such meetings the latest kinds of machines should be described, and the merits of whatever was new should be canvassed. At such meetings projects of a reasonable and practical kind for every improvement should be brought under notice. At such meetings a special report might from time to time be brought up indicating in what directions developments of sound advances and improvements had progressed, or conveying what was being attempted and effected in different parts of the world.

The outline of work here suggested gave scope, I believed, for one of the most active and varied societies, and afforded an opening for talent and industry, through what might be called a pleasure or pastime, which need not, I thought, be described in any but the most plain and unaffected language to win its way. Ladies as well as gentlemen could take a share in adding to the pleasant and useful knowledge which would be collected by the new society. The antiquarians, owing to the rapidity with which they could move from place to place, would very quickly bring into the records of the institution a description of many things of historical interest in these islands. They could furnish reports of every battlefield as it at present exists, every monument, every ruin, every ancient house. They could afford a still more important service. Throughout the country there are scattered in various local museums and libraries, public and private, an immense number of written and printed treasures of which the best scholars have no information, and which, hunted up by members of the energetic cycling fraternity who were devoted to such inquiry, would quickly yield a rich harvest of results. There would, I supposed, be no lack of recruits in this service if the pursuit were but commenced; for, as a matter of experience, we know that researches of this nature soon become fascinating to an extreme degree. One or two good 'finds' of a book or manuscript would stimulate to more exertion, and would reflect such literary credit on the fortunate finder, that the desire to discover would rather have to be kept in bounds than to be eagerly encouraged; because nearly all men and most women love the perusal of the past, and if they have the slightest pretence to culture always listen to history when it is new to them and connected with events affecting the living interests and actions of mankind.

The geographers, who would form the second section of the society, would—I presumed—be equally well placed in regard to the work they might accomplish. There is not a feature in our islands relating to the surface, the culture, the distinctions of counties, not to name other countries, which would not afford them opportunities of observation and record. On one subject alone they might perform a service which to the geographical survey itself would well form a useful supplement, and which could not fail to be appreciated by all governing administrative bodies, local and central. They might in a few months make a report on the state of the roadways, public or private, the highways and byways of the country, the like of which has never existed. They could illustrate what are the best and what are the worst roads; what are safe, what dangerous. They could show what sideways and narrow roads ought to be made for the sake of saving distances when travelling from the great highways to the surrounding towns and villages. They could supply excellent information as to the present state of the old turnpike and the much older Roman roads. They could prepare an admirable chapter on old milestones, and on a number of stone records which have been planted to mark out distances or local peculiarities. By means of their cyclometers they could correct an immense number of errors, which the 'signposts' almost invariably make respecting distances. They could indicate how many 'drift roads' there are in the islands, where these are situated, and what is their extent. They could show what is the best basis of a road, and what upper dressing is most even, most dry and most enduring. They could suggest the parts where, owing to excessive steepness of ascent or descent of surface, it would be proper to set the engineer to work to improve the course by levelling it or diverting the line. They could gather up the very best information on the matter of waste wayside lands and moors. They could map out the parts of the country which are still unenclosed, and record the local usages common in such places for marking out boundaries and for defining public and private rights of property. They could define the lines of rivers and canals, and in so doing could perform what might in future become a most essential service to the commonwealth, by showing where in the course of flowing streams best and steadiest force could be supplied for working the dynamos which must some day be set up on a large scale for producing heat and light for human habitations. They could show what are the best sources of fish-supply, and the readiest means by which stores of fish could be conveyed to the great centres of population. They could discover what parts of the land were most productive of different fruits and vegetables, and indicate the encouragement that ought to be given for the development or sustainment of those cultivations. They could observe what lands were out of cultivation, and what necessity did or did not exist for loss of culture in waste places. They could do all this in reference to our own country; and if they extended their expeditions, as they would be sure to do, to foreign countries—to France, to Spain, to Germany, to Greece, to Italy, and other parts of the continent of Europe, and to other continents—they would become the carriers of as sound and valuable information as any messengers of civilisation and progress who have at any time gone forth to explore and report.

Y

I dwelt at length on the possible work of this section of the proposed society of cyclists because it served conveniently for illustrating the plan that was submitted in its full extent and bearing. I dwelt on it also because it would probably be at first the most popular, as it would certainly be the readiest, section of work. At this moment there are at least a thousand intelligent cyclists, all furnished with machines, all ready equipped for the adventure, all capable of travelling their twenty-five to fifty miles a day, and each one fully capable of adding something to the general stock of information on the topics referred to. If such a corps were duly organised, and if competition went ever so little in comparison with competition for the present 'record,' then in twenty years England might possess from these inquirers a new library of the world, making the planet so well known to our people and other people that all scholars would look upon us as the final conquerors of the geographical sphere—a pre-eminence which England, the parent of the cyclist art and skill, ought assuredly to seize and retain.

There would be equal scope for work in the two other departments which were suggested as parts of the constitution of the new society, namely, natural science, especially natural history and mechanics. Those ladies and gentlemen who are fond of natural history in the various branches to which attention has been drawn would very quickly become mistresses and masters of numerous facts, out of which essays and papers of perfectly original character would be produced. The study of clouds and of weather portents; of dialects of different countries and tracts of country; of prehistoric implements; of remains of extinct animals; of various peculiarities of race; with other subjects allied in character, would be at command, and would yield profitable results to the collectors themselves, as well as to those to whom the knowledge they had acquired would be imparted.

Those, again, who were earnest in following up the improvements of a mechanical kind in cycling machines would find constant opportunity for testing the quality and applicability of assumed advancements, with or without the test now almost entirely relied upon, the long or short race on the track or the road.

To make the new society complete in its working, one special event in its history each year would be the holding of a general conference, with an exhibition of new and improved machines and appliances for cycling purposes. The conference, held in the spring months in one of our large towns in England, Scotland, or Ireland, or even in some continental town, would be, if properly carried out, as good a conference as any of the other great meetings to which the older members of society are wont to repair. It would bring together youth as well as mature age for discussion of subjects the advancement of which are both agreeable to the mind and advantageous to the community. Its organisation, framed on the model of the parent society, would give times for meetings, during which, in the divisions of archæology, geography, natural history, and applied mechanics, there would be opportunity for lively debate, and for comparison from year to year of the progress that was being accomplished in all departments, not excluding a final series of contests for records of the old type, in which 'the sport' need not for any reason be forgotten.

This design, at first sight difficult and improbable, would, I felt confident, be ere long carried out from necessity, unless the art which should give origin to it were permitted to become a mere racing amusement, passing, naturally, and as a matter of course, into disfavour, and falling into the hands of a limited clique bent on keeping it in force for sport alone, and making it thereby unpopular amongst the sedate and intellectual classes of society, whose views and opinions always rule the majority in the long run.

The difficulty of carrying out the design was shown to be at the beginning. The first question that had to be answered was, how shall the start be made. Should some one of the present existing clubs or unions or associations take the lead, and by a modification of its rules be transformed into the new body? Or should there be a new society altogether, starting from its origin with the several objects which had been projected? Or should there be a combination of the existing organisations; and from a central council, formed by them in union, should a new association be founded?

I did not think it signified materially which of the three courses were taken. I believed that the last-named—a combined council of existing bodies—would be the most likely to succeed rapidly. But there were so many obstacles in the way of making a commencement on this basis, and so many contending interests to be taken into account, that probably a new society, starting entirely on new foundations, would be accepted as the fittest mode of development, notwithstanding a certain slowness of growth in the early stages. My task was accomplished in throwing out the idea and in formulating the programme which seemed at the moment to promise the richest results.

THE PROJECT IN FULFILMENT.

In the year 1885 the projected scheme of a society on the lines laid down above was brought into existence, and under the name suggested, 'The Society of Cyclists,' has been in operation for nearly two years. Founded on a new basis, the society has had to contend with all the difficulties which were foretold for it, but in so far as it has progressed it has proved of the greatest benefit to the mental as well as the physical health of those who have taken part in its development. Papers of a valuable kind in all the four subjects, antiquities, natural history, geography, and mechanics, have been read and discussed at the meetings and excursions, and have afforded great pleasure as well as instruction. On one of these occasions a paper read by the honorary secretary of the society, Mr. A. Wynter Blyth, on the subject of 'Cycling in relation to health and beauty,' led to a long and interesting debate bearing on the influence of cycling upon the shape and condition of the body. In this discussion a member of the society, Major Knox-Holmes, who, although he has passed his eightieth year, is still a skilful and enthusiastic rider, submitted many details of evidence of the good effect of the exercise on persons who, like himself, had approached to what are commonly thought to be the confines of a healthy and enduring human existence. His experiences went clearly to show that persons of his own age may, if they are prudent in the way they go to work, receive quite a renewed term of pleasurable activity. He had himself ridden from London to Northampton in one night's ride, and felt no bad effects whatever from the effort.

PRACTICAL HINTS FOR HEALTH-SEEKING TOURISTS.

Let me before I bring this essay to a close dwell very briefly on the points of practice that should always be observed by those persons of sedate natures who really wish to make cycling a healthy exercise for the mind as well as the body.

I.

Let such learners of the art be careful not to carry their first attempts to an extreme. I recommended in one of the earliest papers I wrote on this subject that the beginner should limit his first rides to three or four miles daily. A much esteemed clerical friend of mine told me as a result of his experience that this recommendation was over-cautious. 'I,' said he, 'got on a tricycle for the first time in my life, rode with a friend straight off the reel twenty-seven miles, and from the work felt no unusual fatigue whatever.' The statement naturally enough led me to think that I had been putting too fine a point on distance, and so I ventured to suggest to persons of the same age as my friend, who was a middle-aged man, to begin with six or eight mile rides until they felt fairly at home in the saddle. The advice was followed in several instances, but it did not turn out well. Some quite exceptional faculty of endurance belonged to my critic, or if that were not the case some extreme error was made as to the distance which he had traversed. In brief, I soon became so dissatisfied with the results of the change that I have gone back to the original suggestion. I again recommend riders to be content, at the beginning, to let a ride of three or four miles be all they try to accomplish. At a second ride they can, if they wish, double the distance, and in a few days they may get up to twenty, thirty, and even forty miles a day, the last-named distance being, according to my view, fully sufficient for a long-sustained effort.

For a tour, six hours' work a day at the rate of six miles an hour is sufficient for a person in good condition. For thorough enjoyment on the tour, this may be brought down to five miles an hour continued for five hours each day.

II.

A similar rule as to first ridings should also be recommended to those who have become practised riders but who have got out of condition from want of practice. A busy man leading an in-door life practises the tricycle exercise through the summer months, and during the autumnal vacation finds that five or six hours' ride is a mere pastime and pleasure. He returns to his business and lays by his machine for the winter months. The bright days of spring, always dangerous and critical days, tempt him to sally forth for a ride. He dashes out, runs over four or five miles, forgetting he has to come back. At last, before what seems to him a moderate ride, he feels unusually tired, and turns homeward. Now his weariness soon increases, and by the time he has got home he is thoroughly pumped out. His muscles are stiff and painful; he feels cramps in the calves of his legs the next day; and, do what he will, nay even if he sleep through all the succeeding night, he is exhausted on the following two or three days. To those, then, who are riders I should say again, recommence as you first commenced, and steadily work up to condition if you mean to make the exercise a profit and a pleasure.

III.

It is assumed by some that if walking exercise be maintained, daily cycling exercise can be picked up again at any time. The idea is quite erroneous. Walking exercise and cycling exercise are two perfectly distinct things in their effects on the body and, indirectly, that is through the body, on the mind. The best walkers are not by any necessity the best cyclists, and certainly the best cyclists are not the best walkers. In the long run some men would cover more ground in a given time on a tricycle than on foot; but the reverse may be equally true. Mr. Weston, the famous pedestrian, assured me that he could do a long tour on foot better than on wheels. He had just gone through the remarkable effort of walking fifty miles a day for one hundred days, and he had completed the task by walking the last fifty miles, from Brighton to London, without once sitting down. He said in reference to that memorable feat that he could not have kept it up so well and steadily through all weathers and on all kinds of roads on either a bicycle or a tricycle, and I am quite inclined to believe that he was right, in so far as he himself was concerned.

We must not, therefore, compare the two exercises together, because they admit of contrasts, and these contrasts should be kept in mind in speaking of the exercises of walking and cycling. They are varied exercises. The muscles called into full activity by the one are not the same as are called into exercise by the other. In walking the great extensor muscles in the front part of the thigh, the muscles which extend the leg upon the thigh, are not called into more than ordinary action, because the body being erect they lift the leg forward as on a swing. But in cycling these muscles are brought into great activity, amounting even to strain, the leg being at an angle to the thigh, and so, as every rider learns at first, they are soon painfully wearied. A little walking exercise, however, will remove this strain, and is even better than rest as a means of relief. After a time, in walking, the great muscles of the calf of the leg, which raise the heel, begin early to experience a fatigue, from which in cycling they are very largely saved. Thus in getting down from the machine to walk, the tired muscles of the forepart of the thigh are partially rested; and as on re-mounting and continuing to ride the muscles of the calf of the leg are rested, the fatigue of the limb is equally balanced. The motion of all muscles engaged in cycling is quicker than in walking; and for both these reasons, fatigue when it does occur in cycling is much more sharply developed, is more determinate, and is more prolonged by the act of cycling than by walking. In plain words, only so much work can be got out of a human body in a given time, and if rapidity is to be the first consideration, rapid exhaustion must be submitted to as the necessary consequence.

I do not mention these points in order to run down cycling against walking as an exercise. It is my wish to indicate only that the two exercises are distinct, and that the training for the one is not the training for the other. At the same time one exercise should not exclude the other, and it is quite possible to keep up the training for both with equal success. In this statement there is no paradox, but there is sound argument. The cyclist who is out on tour does well to walk at intervals during the journey, for by that means he not only brings a new set of muscles into play, but he maintains intact the natural muscular activity, so that when he lies down to rest the muscular mechanism is equally exhausted, a condition which is followed by far more equable and refreshing sleep than if one set of muscular organs were utterly worn out, and other sets were under no weariness, or under so little as to require a minimum of rest. The jerks in the limbs, the frequent desire to turn over in bed, the unequal refreshment in the morning which many cyclists complain of after a previous day of heavy riding are due simply to irregular distribution of muscular exhaustion.

So in a tour of thirty-six miles in a day it is healthier to walk one-sixth of the distance, six miles, than to ride the whole distance, even if the road be fairly level. And here is another very good reason for walking up hills instead of climbing them, with infinite labour, on the machine, since in walking and pushing the machine, new sets of muscles are brought into activity, and, as explained above, the equilibrium of muscular work is secured.

IV.

A word was said, in the second paragraph of this *résumé*, about cycling in the spring months after the winter's rest. There is opened up in that reflection a rather important question on the subject of resting from riding and of resuming it. On the whole, it is best for health to keep up moderate cycling, where the practice has once been thoroughly established, and to lose no available opportunity of sustaining the practice in all seasons. What are called regular riders are most favoured riders. Of this I cannot entertain a doubt, and I would add that it is better not to ride at all than to ride excessively in fits and starts. In cases when a rider has to give up the exercise for a long time, he must resume it always with caution, and as it is usually in the spring months that he re-opens his campaign, he should then be more than usually careful in all that he does. Why this should be so is explained in the essay on Dress in Relation to Health. The spring months are of all others the months of waste and feebleness. It is not merely that the weather is most treacherous, but that the body is least able to undergo fatigue and to undertake exhausting trials. Precaution is also necessary now in respect to dress. If the dress be insufficient—and in the transitory warmth of the days of early spring there is the greatest temptation to go out with insufficient dress—the body is liable to exposure to sudden falls of temperature, of an extreme degree, when it is warmest and just when it is becoming fatigued. Under these untoward circumstances the danger of sudden chill and cold is always imminent, and I have known disastrous results to accrue from this kind of exposure.

The lessons derivable from the above observations are: 1. That after a long rest from it, cycling should always be resumed with care and moderation, so as to train towards longer and harder exercise. 2. That when the weather is unsettled and great extremes of heat and cold are apt to replace each other suddenly, provision should be made for carrying an extra covering of clothing to meet the transition from heat to cold.

In order to keep up the muscular training of the cyclist during extremely hard weather, some use at home and indoors what is called the 'home trainer.' This is a fixed machine on which the rider can mount, and by which he can go through any degree of what may be called dummy riding at any speed and under any resistance. I have had one of these machines in my dressing-room for three years, and I find it is really a very good help. They who have not a 'home trainer,' but who have a tricycle, may turn their tricycle into a home trainer by having it raised upon two supports so that its wheels, just clear of the floor, can be made to revolve by working them from the pedals as in riding ordinarily. Then the brake being put on so as to get resistance, the exercise is as good as riding itself, less the advantage of the fresh open air. The distance worked over can be measured by the cyclometer as if the journey were being made on the road.

v.

On alighting from a long ride after free action of the skin, it is proper practice for health's sake to get a change of clothing as quickly as possible after a good cleansing of the skin. Most riders effect the cleansing process by free ablution from head to foot, or, if they can get it, by taking a plunge bath. I am not quite sure whether this is quite the best practice. It seems to me to be always a good plan to subject the skin, in the first instance at any rate, to a good rubbing down with a rough dry towel, and to follow this up with a brisk application of the flesh-brush. After dry scrubbing is actively and effectively done, it is quite sufficient to give the body a sponge over with clean water and dry it again thoroughly with the towel. A bath is a luxury in place of sponging, but is not necessary. The simpler method is equally as good, and it has the great advantage that it can be carried out everywhere. There is no respectable country inn, however simple, that will not afford all the requirements for this method of outward purification, with the exception of the flesh-brush, which the tourist should carry as systematically as he carries his hair-brush. The flesh-brush should not be particularly hard; for a soft brush cleanses best.

After the ablution and rehabiliment in clean dry clothes, a short period of walking exercise is good previous to taking a meal.

VI.

The times of the day that are most fitted for riding, a point often asked, vary with the season. During summer and the early part of autumn, there cannot be much difference of opinion on the subject. The early morning and the late afternoon are unquestionably the two best periods. To be precise I should say, selecting six hours for the touring work of the day, that three hours immediately after sunrise and three hours before sunset form the very best working times. At the end of the first period the rider finds himself ready for a good breakfast, and has then the whole of a long day in his possession for resting, dining, making local excursions in the place he may be visiting, and an hour or two, if he likes, for writing or note-taking. At the end of the second period he finds himself equally ready, after ablution, for a light supper, as preliminary to a sound night's rest. If, however, he be a lazy cyclist, and prefers to remain in bed later than the sun, the next best plan is to get an early breakfast, start on the journey at eight in the morning, continue till ten, then take the day's rest or amusement, recommence the further journey three hours before sunset, and run on an hour after sunset. At times when the moon supplements well the light of the sun, a run of an hour or two is one of the most luxurious enjoyments. For the moment all idea of fatigue seems to vanish, and the knowledge that the journey is about to cease is accompanied with an actual sense of regret. The temptation to go on is indeed too seductive, and requires to be guarded against in order to save the strength of the body for the work of the coming day. The extremest fatigue I ever underwent myself was induced by the seductiveness of a moonlight ride. The muscles being in full activity, and the pleasure of the motion overcoming the idea of weariness, I and my companions sped along for three or four hours at quick pace, after a previous fair day's work. I got off the machine feeling, at the moment, as free from fatigue as when I first mounted it; but this impression was very deceptive. Before going to bed my exhausted muscles were unduly stiffening, I was too wearied to sleep, and was not good for further exercise during the three following days.

VII.

To those cyclists who wish to combine literary work with their exercise I may give a word of information from direct experience which will, I conceive, be a source of pleasure and satisfaction. The exercise of cycling goes excellently well with the exercise of the pen, presuming always that the two exertions, the mental and the physical, are properly balanced, and that neither the one nor the other is carried out to extremes. Cycling exercise, moderately employed and combined with the change of air which is connected with it, makes the mind extremely buoyant, refreshes the memory, and renders literary labour peculiarly pleasant, rapid, and steady. But just the same precaution in reference to literary exertion as that which has been mentioned in relation to the labour of cycling itself is required. The ease with which composition is carried on for a short time is seductive towards undue continuance of it, and where from indulgence in the temptation the mind becomes imperceptibly weary, the body as imperceptibly falls in turn into the train of weariness. I have made here a physiological observation which is very curious simply as an observation, and is useful in the light of practical experience. The observation is this: that mental exhaustion leads much more quickly to physical exhaustion than physical leads to mental. From moderate physical exhaustion, presuming that the mental faculties have not been imposed upon too severely, recuperation is very rapid and complete; but if the mind has been previously tired the same degree of physical exhaustion leads to a much more prolonged state of comparative helplessness both of body and of mind. For now the fatigue of body keeps up the fatigue of mind, and the whole organisation, mental and physical, must rest in order to obtain perfected, re-created activity.

THE END.

PRINTED BY
SPOTTISWOODE AND CO., NEW-STREET SQUARE
LONDON

MARCH 1887.

GENERAL LISTS OF WORKS
PUBLISHED BY
MESSRS. LONGMANS, GREEN, & CO.
39 PATERNOSTER ROW, LONDON, E.C.

HISTORY, POLITICS, HISTORICAL MEMOIRS, &c.

Abbey's The English Church and its Bishops, 1700-1800. 2 vols. 8vo. 24s.
Abbey and Overton's English Church in the Eighteenth Century. Cr. 8vo. 7s. 6d.
Arnold's Lectures on Modern History. 8vo. 7s. 6d.
Bagwell's Ireland under the Tudors. Vols. 1 and 2. 2 vols. 8vo. 32s.
Ball's The Reformed Church of Ireland, 1537-1886. 8vo. 7s. 6d.
Boultbee's History of the Church of England, Pre-Reformation Period. 8vo. 15s.
Buckle's History of Civilisation. 3 vols. crown 8vo. 24s.
Cox's (Sir G. W.) General History of Greece. Crown 8vo. Maps, 7s. 6d.
Creighton's History of the Papacy during the Reformation. 8vo. Vols. 1 and 2, 32s. Vols. 3 and 4, 24s.
De Tocqueville's Democracy in America. 2 vols. crown 8vo. 16s.
Doyle's English in America : Virginia, Maryland, and the Carolinas, 8vo. 18s.
  —   —   — The Puritan Colonies, 2 vols. 8vo. 36s.
Epochs of Ancient History. Edited by the Rev. Sir G. W. Cox, Bart. and C. Sankey, M.A. With Maps. Fcp. 8vo. price 2s. 6d. each.

Beesly's Gracchi, Marius, and Sulla.
Capes's Age of the Antonines.
  — Early Roman Empire.
Cox's Athenian Empire.
  — Greeks and Persians.
Curteis's Rise of the Macedonian Empire.

Ihne's Rome to its Capture by the Gauls.
Merivale's Roman Triumvirates.
Sankey's Spartan and Theban Supremacies.
Smith's Rome and Carthage, the Punic Wars.

Epochs of Modern History. Edited by C. Colbeck, M.A. With Maps. Fcp. 8vo. price 2s. 6d. each.

Church's Beginning of the Middle Ages.
Cox's Crusades.
Creighton's Age of Elizabeth.
Gairdner's Houses of Lancaster and York.
Gardiner's Puritan Revolution.
  — Thirty Years' War.
  — (Mrs.) French Revolution, 1789-1795.
Hale's Fall of the Stuarts.
Johnson's Normans in Europe.

Longman's Frederick the Great and the Seven Years' War.
Ludlow's War of American Independence.
M'Carthy's Epoch of Reform, 1830-1850.
Moberly's The Early Tudors.
Morris's Age o' Queen Anne.
  — The Early Hanoverians.
Seebohm's Protestant Revolution.
Stubbs's The Early Plantagenets.
Warburton's Edward III.

Epochs of Church History. Edited by the Rev. Mandell Creighton, M.A. Fcp. 8vo. price 2s. 6d. each.

Brodrick's A History of the University of Oxford.
Overton's The Evangelical Revival in the Eighteenth Century.

Perry's The Reformation in England.
Plummer's The Church of the Early Fathers.
Tucker's The English Church in other Lands.

_{*} *Other Volumes in preparation.*

London : LONGMANS, GREEN, & CO.

Freeman's Historical Geography of Europe. 2 vols. 8vo. 31s. 6d.
Froude's English in Ireland in the 18th Century. 3 vols. crown 8vo. 18s.
— History of England. Popular Edition. 12 vols. crown 8vo. 3s. 6d. each.
Gardiner's History of England from the Accession of James I. to the Outbreak of the Civil War. 10 vols. crown 8vo. 60s.
— History of the Great Civil War, 1642-1649 (3 vols.) Vol. 1, 1642-1644, 8vo. 21s.
Greville's Journal of the Reign of Queen Victoria, 1837-1852. 3 vols. 8vo. 36s. 1852-1860, 2 vols. 8vo. 24s.
Historic Towns. Edited by E. A. Freeman, D.C.L. and Rev. William Hunt, M.A. With Maps and Plans. Crown 8vo. 3s. 6d. each.

| London. By W. E. Loftie. | Bristol. By Rev. W. Hunt. |
| Exeter. By E. A. Freeman. | |

*** *Other volumes in preparation.*

Lecky's History of England in the Eighteenth Century. Vols. 1 & 2, 1700-1760, 8vo. 36s. Vols. 3 & 4, 1760-1784, 8vo. 36s.
— History of European Morals. 2 vols. crown 8vo. 16s.
— — — Rationalism in Europe. 2 vols. crown 8vo. 16s.
Longman's Life and Times of Edward III. 3 vols. 8vo. 28s.
Macaulay's Complete Works. Library Edition. 8 vols. 8vo. £5. 5s.
— — Cabinet Edition. 16 vols. crown 8vo. £4. 16s.
— History of England :—

| Student's Edition. 2 vols. cr. 8vo. 12s. | Cabinet Edition. 8 vols. post 8vo. 48s. |
| People's Edition. 4 vols. cr. 8vo. 16s. | Library Edition. 5 vols. 8vo. £4. |

Macaulay's Critical and Historical Essays, with Lays of Ancient Rome In One Volume :—

| Authorised Edition. Cr. 8vo. 2s. 6d. or 3s. 6d. gilt edges. | Popular Edition. Cr. 8vo. 2s. 6d. |

Macaulay's Critical and Historical Essays :—

| Student's Edition. 1 vol. cr. 8vo. 6s. | Cabinet Edition. 4 vols. post 8vo. 24s. |
| People's Edition. 2 vols. cr. 8vo. 8s. | Library Edition. 3 vols. 8vo. 36s. |

Macaulay's Speeches corrected by Himself. Crown 8vo. 3s. 6d.
Malmesbury's (Earl of) Memoirs of an Ex-Minister. Crown 8vo. 7s. 6d.
Maxwell's (Sir W. S.) Don John of Austria. Library Edition, with numerous Illustrations. 2 vols. royal 8vo. 42s.
May's Constitutional History of England, 1760-1870. 3 vols. crown 8vo. 18s.
— Democracy in Europe. 2 vols. 8vo. 32s.
Merivale's Fall of the Roman Republic. 12mo. 7s. 6d.
— General History of Rome, B.C. 753-A.D. 476. Crown 8vo. 7s. 6d.
— History of the Romans under the Empire. 8 vols. post 8vo. 48s.
Nelson's (Lord) Letters and Despatches. Edited by J. K. Laughton. 8vo. 16s.
Outlines of Jewish History from B.C. 586 to C.E. 1885. By the author of 'About the Jews since Bible Times.' Fcp. 8vo. 3s. 6d.
Pears' The Fall of Constantinople. 8vo. 16s.
Seebohm's Oxford Reformers—Colet, Erasmus, & More. 8vo. 14s.
Short's History of the Church of England. Crown 8vo. 7s. 6d.
Smith's Carthage and the Carthaginians. Crown 8vo. 10s. 6d.
Taylor's Manual of the History of India. Crown 8vo. 7s. 6d.

London: LONGMANS, GREEN, & CO.

Walpole's History of England, from 1815. 5 vols. 8vo. Vols. 1 & 2, 1815–1832, 36s. Vol. 3, 1832–1841, 18s. Vols. 4 & 5, 1841–1858, 36s.
Wylie's History of England under Henry IV. Vol. 1, crown 8vo. 10s. 6d.

BIOGRAPHICAL WORKS.

Armstrong's (E. J.) Life and Letters. Edited by G. F. Armstrong. Fcp. 8vo. 7s. 6d.
Bacon's Life and Letters, by Spedding. 7 vols. 8vo. £4. 4s.
Bagehot's Biographical Studies. 1 vol. 8vo. 12s.
Carlyle's Life, by J. A. Froude. Vols. 1 & 2, 1795–1835, 8vo. 32s. Vols. 3 & 4, 1834–1881, 8vo. 32s.
— (Mrs.) Letters and Memorials. 3 vols. 8vo. 36s.
Doyle (Sir F. H.) Reminiscences and Opinions. 8vo. 16s.
English Worthies. Edited by Andrew Lang. Crown 8vo. 2s. 6d. each.
  Charles Darwin. By Grant Allen. | Marlborough. By George Saintsbury.
  Shaftesbury (The First Earl). By | Steele. By Austin Dobson.
  H. D. Traill. | Ben Jonson. By J. A. Symonds.
  Admiral Blake. By David Hannay. | George Canning. By Frank H. Hill.
  *** *Other Volumes in preparation.*
Fox (Charles James) The Early History of. By Sir G. O. Trevelyan, Bart. Crown 8vo. 6s.
Froude's Cæsar: a Sketch. Crown 8vo. 6s.
Hamilton's (Sir W. R.) Life, by Graves. Vols. 1 and 2, 8vo. 15s. each.
Havelock's Life, by Marshman. Crown 8vo. 3s. 6d.
Hobart Pacha's Sketches from my Life. Crown 8vo. 7s. 6d.
Macaulay's (Lord) Life and Letters. By his Nephew, Sir G. O. Trevelyan, Bart. Popular Edition, 1 vol. crown 8vo. 6s. Cabinet Edition, 2 vols. post 8vo. 12s. Library Edition, 2 vols. 8vo. 36s.
Mendelssohn's Letters. Translated by Lady Wallace. 2 vols. cr. 8vo. 5s. each.
Mill (James) Biography of, by Prof. Bain. Crown 8vo. 5s.
— (John Stuart) Recollections of, by Prof. Bain. Crown 8vo. 2s. 8d.
— Autobiography. 8vo. 7s. 6d.
Müller's (Max) Biographical Essays. Crown 8vo. 7s. 6d.
Newman's Apologia pro Vitâ Suâ. Crown 8vo. 6s.
Pasteur (Louis) His Life and Labours. Crown 8vo. 7s. 6d.
Shakespeare's Life (Outlines of), by Halliwell-Phillipps. 2 vols. royal 8vo. 10s. 6d.
Southey's Correspondence with Caroline Bowles. 8vo. 14s.
Stephen's Essays in Ecclesiastical Biography. Crown 8vo. 7s. 6d.
Wellington's Life, by Gleig. Crown 8vo. 6s.

MENTAL AND POLITICAL PHILOSOPHY, FINANCE, &c.

Amos's View of the Science of Jurisprudence. 8vo. 18s.
— Primer of the English Constitution. Crown 8vo. 6s.
Bacon's Essays, with Annotations by Whately. 8vo. 10s. 6d.
— Works, edited by Spedding. 7 vols. 8vo. 73s. 6d.
Bagehot's Economic Studies, edited by Hutton. 8vo. 10s. 6d.
— The Postulates of English Political Economy. Crown 8vo. 2s. 6d.
Bain's Logic, Deductive and Inductive. Crown 8vo. 10s. 6d.
  PART I. Deduction, 4s. | PART II. Induction, 6s. 6d.
— Mental and Moral Science. Crown 8vo. 10s. 6d.
— The Senses and the Intellect. 8vo. 15s.
— The Emotions and the Will. 8vo. 15s.
— Practical Essays. Crown 8vo. 4s. 6d.

London: LONGMANS, GREEN, & CO.

General Lists of Works.

Buckle's (H. T.) Miscellaneous and Posthumous Works. 2 vols. crown 8vo. 21s.
Crozier's Civilization and Progress. 8vo. 14s.
Crump's A Short Enquiry into the Formation of English Political Opinion. 8vo. 7s. 6d.
Dowell's A History of Taxation and Taxes in England. 4 vols. 8vo. 48s.
Green's (Thomas Hill) Works. (3 vols.) Vols. 1 & 2, Philosophical Works. 8vo. 16s. each.
Hume's Essays, edited by Green & Grose. 2 vols. 8vo. 28s.
— Treatise of Human Nature, edited by Green & Grose. 2 vols. 8vo. 28s.
Lang's Custom and Myth : Studies of Early Usage and Belief. Crown 8vo. 7s. 6d.
Leslie's Essays in Political and Moral Philosophy. 8vo. 10s. 6d.
Lewes's History of Philosophy. 2 vols. 8vo. 32s.
Lubbock's Origin of Civilisation. 8vo. 18s.
Macleod's Principles of Economical Philosophy. In 2 vols. Vol. 1, 8vo. 15s. Vol. 2, Part I. 12s.
— The Elements of Economics. (2 vols.) Vol. 1, cr. 8vo. 7s. 6d. Vol. 2, Part I. cr. 8vo. 7s. 6d.
— The Elements of Banking. Crown 8vo. 5s.
— The Theory and Practice of Banking. Vol. 1, 8vo. 12s. Vol. 2, 14s.
— Economics for Beginners. 8vo. 2s. 6d.
— Lectures on Credit and Banking. 8vo. 5s.
Mill's (James) Analysis of the Phenomena of the Human Mind. 2 vols. 8vo. 28s.
Mill (John Stuart) on Representative Government. Crown 8vo. 2s.
— — on Liberty. Crown 8vo. 1s. 4d.
— — Examination of Hamilton's Philosophy. 8vo. 16s.
— — Logic. Crown 8vo. 5s.
— — Principles of Political Economy. 2 vols. 8vo. 30s. People's Edition, 1 vol. crown 8vo. 5s.
— — Subjection of Women. Crown 8vo. 6s.
— — Utilitarianism. 8vo. 5s.
— — Three Essays on Religion, &c. 8vo. 5s.
Mulhall's History of Prices since 1850. Crown 8vo. 6s.
Sandars's Institutes of Justinian, with English Notes. 8vo. 18s.
Seebohm's English Village Community. 8vo. 16s.
Sully's Outlines of Psychology. 8vo. 12s. 6d.
— Teacher's Handbook of Psychology. Crown 8vo. 6s. 6d.
Swinburne's Picture Logic. Post 8vo. 5s.
Thompson's A System of Psychology. 2 vols. 8vo. 36s.
Thomson's Outline of Necessary Laws of Thought. Crown 8vo. 6s.
Twiss's Law of Nations in Time of War. 8vo. 21s.
— — in Time of Peace. 8vo. 15s.
Webb's The Veil of Isis. 8vo. 10s. 6d.
Whately's Elements of Logic. Crown 8vo. 4s. 6d.
— — — Rhetoric. Crown 8vo. 4s. 6d.
Wylie's Labour, Leisure, and Luxury. Crown 8vo. 6s.
Zeller's History of Eclecticism in Greek Philosophy. Crown 8vo. 10s. 6d.
— Plato and the Older Academy. Crown 8vo. 18s.
— Pre-Socratic Schools. 2 vols. crown 8vo. 30s.
— Socrates and the Socratic Schools. Crown 8vo. 10s. 6d.
— Stoics, Epicureans, and Sceptics. Crown 8vo. 15s.
— Outlines of the History of Greek Philosophy. Crown 8vo. 10s. 6d.

London : LONGMANS, GREEN, & CO.

MISCELLANEOUS WORKS.

A. K. H. B., The Essays and Contributions of. Crown 8vo.
    Autumn Holidays of a Country Parson. 3s. 6d.
    Changed Aspects of Unchanged Truths. 3s. 6d.
    Common-Place Philosopher in Town and Country. 3s. 6d.
    Critical Essays of a Country Parson. 3s. 6d.
    Counsel and Comfort spoken from a City Pulpit. 3s. 6d.
    Graver Thoughts of a Country Parson. Three Series. 3s. 6d. each.
    Landscapes, Churches, and Moralities. 3s. 6d.
    Leisure Hours in Town. 3s. 6d. Lessons of Middle Age. 3s. 6d.
    Our Homely Comedy; and Tragedy. 3s. 6d.
    Our Little Life. Essays Consolatory and Domestic. Two Series. 3s. 6d. [each.
    Present-day Thoughts. 3s. 6d.
    Recreations of a Country Parson. Three Series. 3s. 6d. each.
    Seaside Musings on Sundays and Week-Days. 3s. 6d.
    Sunday Afternoons in the Parish Church of a University City. 3s. 6d.
Armstrong's (Ed. J.) Essays and Sketches. Fcp. 8vo. 5s.
Arnold's (Dr. Thomas) Miscellaneous Works. 8vo. 7s. 6d.
Bagehot's Literary Studies, edited by Hutton. 2 vols. 8vo. 28s.
Beaconsfield (Lord), The Wit and Wisdom of. Crown 8vo. 1s. boards; 1s. 6d. cl.
Evans's Bronze Implements of Great Britain. 8vo. 25s.
Farrar's Language and Languages. Crown 8vo. 6s.
Froude's Short Studies on Great Subjects. 4 vols. crown 8vo. 24s.
Lang's Letters to Dead Authors. Fcp. 8vo. 6s. 6d.
   — Books and Bookmen. Crown 8vo. 6s. 6d.
Macaulay's Miscellaneous Writings. 2 vols. 8vo. 21s. 1 vol. crown 8vo. 4s. 6d.
   — Miscellaneous Writings and Speeches. Crown 8vo. 6s.
   — Miscellaneous Writings, Speeches, Lays of Ancient Rome, &c. Cabinet Edition. 4 vols. crown 8vo. 24s.
   — Writings, Selections from. Crown 8vo. 6s.
Müller's (Max) Lectures on the Science of Language. 2 vols. crown 8vo. 16s.
   — — Lectures on India. 8vo. 12s. 6d.
Proctor's Chance and Luck. Crown 8vo. 5s.
Smith (Sydney) The Wit and Wisdom of. Crown 8vo. 1s. boards; 1s. 6d. cloth.

ASTRONOMY.

Herschel's Outlines of Astronomy. Square crown 8vo. 12s.
Proctor's Larger Star Atlas. Folio, 15s. or Maps only, 12s. 6d.
   — New Star Atlas. Crown 8vo. 5s.
   — Light Science for Leisure Hours. 3 Series. Crown 8vo. 5s. each.
   — The Moon. Crown 8vo. 6s.
   — Other Worlds than Ours. Crown 8vo. 5s.
   — The Sun. Crown 8vo. 14s.
   — Studies of Venus-Transits. 8vo. 5s.
   — Orbs Around Us. Crown 8vo. 5s.
   — Universe of Stars. 8vo. 10s. 6d.
Webb's Celestial Objects for Common Telescopes. Crown 8vo. 9s.

THE 'KNOWLEDGE' LIBRARY.
Edited by RICHARD A. PROCTOR.

How to Play Whist. Crown 8vo. 5s.
Home Whist. 16mo. 1s.
The Borderland of Science. Cr. 8vo. 6s.
Nature Studies. Crown 8vo. 6s.
Leisure Readings. Crown 8vo. 6s.
The Stars in their Seasons. Imp. 8vo. 5s.
Myths and Marvels of Astronomy. Crown 8vo. 6s.

Pleasant Ways in Science. Cr. 8vo. 6s.
Star Primer. Crown 4to. 2s. 6d.
The Seasons Pictured. Demy 4to. 5s.
Strength and Happiness. Cr. 8vo. 5s.
Rough Ways made Smooth. Cr. 8vo. 6s.
The Expanse of Heaven. Cr. 8vo. 5s.
Our Place among Infinities. Cr. 8vo. 5s.

London: LONGMANS, GREEN, & CO.

CLASSICAL LANGUAGES AND LITERATURE.

Æschylus, The Eumenides of. Text, with Metrical English Translation, by J. F. Davies. 8vo. 7s.
Aristophanes' The Acharnians, translated by R. Y. Tyrrell. Crown 8vo. 2s. 6d.
Aristotle's The Ethics, Text and Notes, by Sir Alex. Grant, Bart. 2 vols. 8vo. 32s.
— The Nicomachean Ethics, translated by Williams, crown 8vo. 7s. 6d.
— The Politics, Books I. III. IV. (VII.) with Translation, &c. by Bolland and Lang. Crown 8vo. 7s. 6d.
Becker's *Charicles* and *Gallus*, by Metcalfe. Post 8vo. 7s. 6d. each.
Cicero's Correspondence, Text and Notes, by R. Y. Tyrrell. Vols. 1 & 2, 8vo. 12s. each.
Homer's Iliad, Homometrically translated by Cayley. 8vo. 12s. 6d.
— — Greek Text, with Verse Translation, by W. C. Green. Vol. 1, Books I.-XII. Crown 8vo. 6s.
Mahaffy's Classical Greek Literature. Crown 8vo. Vol. 1, The Poets, 7s. 6d. Vol. 2, The Prose Writers, 7s. 6d.
Plato's Parmenides, with Notes, &c. by J. Maguire. 8vo. 7s. 6d.
Virgil's Works, Latin Text, with Commentary, by Kennedy. Crown 8vo. 10s. 6d.
— Æneid, translated into English Verse, by Conington. Crown 8vo. 9s.
— — — — — by W. J. Thornhill. Cr. 8vo. 7s.6d.
— Poems, — — — Prose, by Conington. Crown 8vo. 9s.
Witt's Myths of Hellas, translated by F. M. Younghusband. Crown 8vo. 3s. 6d.
— The Trojan War, — — Fcp. 8vo. 2s.
— The Wanderings of Ulysses, — Crown 8vo. 3s. 6d.

NATURAL HISTORY, BOTANY, & GARDENING.

Allen's Flowers and their Pedigrees. Crown 8vo. Woodcuts, 5s.
Decaisne and Le Maout's General System of Botany. Imperial 8vo. 31s. 6d.
Dixon's Rural Bird Life. Crown 8vo. Illustrations, 5s.
Hartwig's Aerial World, 8vo. 10s. 6d.
— Polar World, 8vo. 10s. 6d.
— Sea and its Living Wonders. 8vo. 10s. 6d.
— Subterranean World, 8vo. 10s. 6d.
— Tropical World, 8vo. 10s. 6d.
Lindley's Treasury of Botany. 2 vols. fcp. 8vo. 12s.
Loudon's Encyclopædia of Gardening. 8vo. 21s.
— — Plants. 8vo. 42s.
Rivers's Orchard House. Crown 8vo. 5s.
— Miniature Fruit Garden. Fcp. 8vo. 4s.
Stanley's Familiar History of British Birds. Crown 8vo. 6s.
Wood's Bible Animals. With 112 Vignettes. 8vo. 10s. 6d.
— Common British Insects. Crown 8vo. 3s. 6d.
— Homes Without Hands, 8vo. 10s. 6d.
— Insects Abroad, 8vo. 10s. 6d.
— Horse and Man. 8vo. 14s.
— Insects at Home. With 700 Illustrations. 8vo. 10s. 6d.
— Out of Doors. Crown 8vo. 5s.
— Petland Revisited. Crown 8vo. 7s. 6d.
— Strange Dwellings. Crown 8vo. 5s. Popular Edition, 4to. 6d.

London: LONGMANS, GREEN, & CO.

THE FINE ARTS AND ILLUSTRATED EDITIONS.

Eastlake's Household Taste in Furniture, &c. Square crown 8vo. 14s.
Jameson's Sacred and Legendary Art. 6 vols. square 8vo.
  Legends of the Madonna. 1 vol. 21s.
  — — — Monastic Orders 1 vol. 21s.
  — — — Saints and Martyrs. 2 vols. 31s. 6d.
  — — — Saviour. Completed by Lady Eastlake. 2 vols. 42s.
Macaulay's Lays of Ancient Rome, illustrated by Scharf. Fcp. 4to. 10s. 6d.
The same, with Ivry and the Armada, illustrated by Weguelin. Crown 8vo. 3s. 6d.
New Testament (The) illustrated with Woodcuts after Paintings by the Early Masters. 4to. 21s.

CHEMISTRY ENGINEERING, & GENERAL SCIENCE.

Arnott's Elements of Physics or Natural Philosophy. Crown 8vo. 12s. 6d.
Barrett's English Glees and Part-Songs: their Historical Development. Crown 8vo. 7s. 6d.
Bourne's Catechism of the Steam Engine. Crown 8vo. 7s. 6d.
  — Examples of Steam, Air, and Gas Engines. 4to. 70s.
  — Handbook of the Steam Engine. Fcp. 8vo. 9s.
  — Recent Improvements in the Steam Engine. Fcp. 8vo. 6s.
  — Treatise on the Steam Engine. 4to. 42s.
Buckton's Our Dwellings, Healthy and Unhealthy. Crown 8vo. 3s. 6d.
Clerk's The Gas Engine. With Illustrations. Crown 8vo. 7s. 6d.
Crookes's Select Methods in Chemical Analysis. 8vo. 24s.
Culley's Handbook of Practical Telegraphy. 8vo. 16s.
Fairbairn's Useful Information for Engineers. 3 vols. crown 8vo. 31s. 6d.
  — Mills and Millwork. 1 vol. 8vo. 25s.
Ganot's Elementary Treatise on Physics, by Atkinson. Large crown 8vo. 15s.
  — Natural Philosophy, by Atkinson. Crown 8vo. 7s. 6d.
Grove's Correlation of Physical Forces. 8vo. 15s.
Haughton's Six Lectures on Physical Geography. 8vo. 15s.
Helmholtz on the Sensations of Tone. Royal 8vo. 28s.
Helmholtz's Lectures on Scientific Subjects. 2 vols. crown 8vo. 7s. 6d. each.
Hudson and Gosse's The Rotifera or 'Wheel Animalcules.' With 30 Coloured Plates. 6 parts. 4to. 10s. 6d. each. Complete, 2 vols. 4to. £3. 10s.
Hullah's Lectures on the History of Modern Music. 8vo. 8s. 6d.
  — Transition Period of Musical History. 8vo. 10s. 6d.
Jackson's Aid to Engineering Solution. Royal 8vo. 21s.
Jago's Inorganic Chemistry, Theoretical and Practical. Fcp. 8vo. 2s.
Jeans' Railway Problems. 8vo. 12s. 6d.
Kolbe's Short Text-Book of Inorganic Chemistry. Crown 8vo. 7s. 6d.
Lloyd's Treatise on Magnetism. 8vo. 10s. 6d.
Macalister's Zoology and Morphology of Vertebrate Animals. 8vo. 10s. 6d.
Macfarren's Lectures on Harmony. 8vo. 12s.
Miller's Elements of Chemistry, Theoretical and Practical. 3 vols. 8vo. Part I. Chemical Physics, 16s. Part II. Inorganic Chemistry, 24s. Part III. Organic Chemistry, price 31s. 6d.
Mitchell's Manual of Practical Assaying. 8vo. 31s. 6d.

London: LONGMANS, GREEN, & CO.

Noble's Hours with a Three-inch Telescope. Crown 8vo. 4s. 6d.
Northcott's Lathes and Turning. 8vo. 18s.
Owen's Comparative Anatomy and Physiology of the Vertebrate Animals. 3 vols. 8vo. 73s. 6d.
Piesse's Art of Perfumery. Square crown 8vo. 21s.
Reynolds's Experimental Chemistry. Fcp. 8vo. Part I. 1s. 6d. Part II. 2s. 6d. Part III. 3s. 6d.
Schellen's Spectrum Analysis. 8vo. 31s. 6d.
Sennett's Treatise on the Marine Steam Engine. 8vo. 21s.
Smith's Air and Rain. 8vo. 24s.
Stoney's The Theory of the Stresses on Girders, &c. Royal 8vo. 36s.
Tilden's Practical Chemistry. Fcp. 8vo. 1s. 6d.
Tyndall's Faraday as a Discoverer. Crown 8vo. 3s. 6d.
— Floating Matter of the Air. Crown 8vo. 7s. 6d.
— Fragments of Science. 2 vols. post 8vo. 16s.
— Heat a Mode of Motion. Crown 8vo. 12s.
— Lectures on Light delivered in America. Crown 8vo. 5s.
— Lessons on Electricity. Crown 8vo. 2s. 6d.
— Notes on Electrical Phenomena. Crown 8vo. 1s. sewed, 1s. 6d. cloth.
— Notes of Lectures on Light. Crown 8vo. 1s. sewed, 1s. 6d. cloth.
— Sound, with Frontispiece and 203 Woodcuts. Crown 8vo. 10s. 6d.
Watts's Dictionary of Chemistry. 9 vols. medium 8vo. £15. 2s. 6d.
Wilson's Manual of Health-Science. Crown 8vo. 2s. 6d.

THEOLOGICAL AND RELIGIOUS WORKS.

Arnold's (Rev. Dr. Thomas) Sermons. 6 vols. crown 8vo. 5s. each.
Boultbee's Commentary on the 39 Articles. Crown 8vo. 6s.
Browne's (Bishop) Exposition of the 39 Articles. 8vo. 16s.
Bullinger's Critical Lexicon and Concordance to the English and Greek New Testament. Royal 8vo. 15s.
Colenso on the Pentateuch and Book of Joshua. Crown 8vo. 6s.
Conder's Handbook of the Bible. Post 8vo. 7s. 6d.
Conybeare & Howson's Life and Letters of St. Paul :—
    Library Edition, with Maps, Plates, and Woodcuts. 2 vols. square crown 8vo. 21s.
    Student's Edition, revised and condensed, with 46 Illustrations and Maps. 1 vol. crown 8vo. 7s. 6d.
Cox's (Homersham) The First Century of Christianity. 8vo. 12s.
Davidson's Introduction to the Study of the New Testament. 2 vols. 8vo. 30s.
Edersheim's Life and Times of Jesus the Messiah. 2 vols. 8vo. 24s.
— Prophecy and History in relation to the Messiah. 8vo. 12s.
Ellicott's (Bishop) Commentary on St. Paul's Epistles. 8vo. Galatians, 8s. 6d. Ephesians, 8s. 6d. Pastoral Epistles, 10s. 6d. Philippians, Colossians and Philemon, 10s. 6d. Thessalonians, 7s. 6d.
— Lectures on the Life of our Lord. 8vo. 12s.
Ewald's Antiquities of Israel, translated by Solly. 8vo. 12s. 6d.
— History of Israel, translated by Carpenter & Smith. 8 vols. 8vo. Vols. 1 & 2, 24s. Vols. 3 & 4, 21s. Vol. 5, 18s. Vol. 6, 16s. Vol. 7, 21s. Vol. 8, 18s.
Hobart's Medical Language of St. Luke. 8vo. 16s.
Hopkins's Christ the Consoler. Fcp. 8vo. 2s. 6d.

London: LONGMANS, GREEN, & CO.

General Lists of Works. 9

Jukes's New Man and the Eternal Life. Crown 8vo. 6s.
— Second Death and the Restitution of all Things. Crown 8vo. 3s. 6d.
— Types of Genesis. Crown 8vo. 7s. 6d.
— The Mystery of the Kingdom. Crown 8vo. 3s. 6d.
Lenormant's New Translation of the Book of Genesis. Translated into English. 8vo. 10s. 6d.
Lyra Germanica: Hymns translated by Miss Winkworth. Fcp. 8vo. 5s.
Macdonald's (G.) Unspoken Sermons. Two Series, Crown 8vo. 3s. 6d. each.
— The Miracles of our Lord. Crown 8vo. 3s. 6d.
Manning's Temporal Mission of the Holy Ghost. Crown 8vo. 8s. 6d.
Martineau's Endeavours after the Christian Life. Crown 8vo. 7s. 6d.
— Hymns of Praise and Prayer. Crown 8vo. 4s. 6d. 32mo. 1s. 6d.
— Sermons, Hours of Thought on Sacred Things. 2 vols. 7s. 6d. each.
Monsell's Spiritual Songs for Sundays and Holidays. Fcp. 8vo. 5s. 18mo. 2s.
Müller's (Max) Origin and Growth of Religion. Crown 8vo. 7s. 6d.
— Science of Religion. Crown 8vo. 7s. 6d.
Newman's Apologia pro Vitâ Suâ. Crown 8vo. 6s.
— The Idea of a University Defined and Illustrated. Crown 8vo. 7s.
— Historical Sketches. 3 vols. crown 8vo. 6s. each.
— Discussions and Arguments on Various Subjects. Crown 8vo. 6s.
— An Essay on the Development of Christian Doctrine. Crown 8vo. 6s.
— Certain Difficulties Felt by Anglicans in Catholic Teaching Considered. Vol. 1, crown 8vo. 7s. 6d. Vol. 2, crown 8vo. 5s. 6d.
— The Via Media of the Anglican Church, Illustrated in Lectures, &c. 2 vols. crown 8vo. 6s. each
— Essays, Critical and Historical. 2 vols. crown 8vo. 12s.
— Essays on Biblical and on Ecclesiastical Miracles. Crown 8vo. 6s.
— An Essay in Aid of a Grammar of Assent. 7s. 6d.
Overton's Life in the English Church (1660-1714). 8vo. 14s.
Supernatural Religion. Complete Edition. 3 vols. 8vo. 36s.
Younghusband's The Story of Our Lord told in Simple Language for Children. Illustrated. Crown 8vo. 2s. 6d. cloth plain; 3s. 6d. cloth extra, gilt edges.

TRAVELS, ADVENTURES, &c.

Alpine Club (The) Map of Switzerland. In Four Sheets. 42s.
Baker's Eight Years in Ceylon. Crown 8vo. 5s.
— Rifle and Hound in Ceylon. Crown 8vo. 5s.
Ball's Alpine Guide. 3 vols. post 8vo. with Maps and Illustrations:—I. Western Alps, 6s. 6d. II. Central Alps, 7s. 6d. III. Eastern Alps, 10s. 6d.
Ball on Alpine Travelling, and on the Geology of the Alps, 1s.
Brassey's Sunshine and Storm in the East. Library Edition, 8vo. 21s. Cabinet Edition, crown 8vo. 7s. 6d. Popular Edition, 4to. 6d.
— Voyage in the Yacht 'Sunbeam.' Library Edition, 8vo. 21s. Cabinet Edition, crown 8vo. 7s. 6d. School Edition, fcp. 8vo. 2s. Popular Edition, 4to. 6d.
— In the Trades, the Tropics, and the 'Roaring Forties.' Library Edition, 8vo. 21s. Cabinet Edition, crown 8vo. 17s. 6d. Popular Edition, 4to. 6d.
Froude's Oceana; or, England and her Colonies. Crown 8vo. 2s. boards; 2s. 6d. cloth.
Howitt's Visits to Remarkable Places. Crown 8vo. 7s. 6d.
Three in Norway. By Two of Them. Crown 8vo. Illustrations, 6s.

London: LONGMANS, GREEN, & CO.

SPORTS AND PASTIMES.

The Badminton Library of Sports and Pastimes. Edited by the Duke of Beaufort and A. E. T. Watson. With numerous Illustrations. Crown 8vo. 10s. 6d. each.
 Hunting, by the Duke of Beaufort, &c.
 Fishing, by H. Cholmondeley-Pennell, &c. 2 vols.
 Racing, by the Earl of Suffolk, &c.
 Shooting, by Lord Walsingham, &c. 2 vols.
 Cycling. By Viscount Bury.
 ⁎ *Other Volumes in preparation.*
Campbell-Walker's Correct Card, or How to Play at Whist. Fcp. 8vo. 2s. 6d.
Dead Shot (The) by Marksman. Crown 8vo. 10s. 6d.
Francis's Treatise on Fishing in all its Branches. Post 8vo. 15s.
Longman's Chess Openings. Fcp. 8vo. 2s. 6d.
Pease's The Cleveland Hounds as a Trencher-Fed Pack. Royal 8vo. 18s.
Pole's Theory of the Modern Scientific Game of Whist. Fcp. 8vo. 2s. 6d.
Proctor's How to Play Whist. Crown 8vo. 5s.
Ronalds's Fly-Fisher's Entomology. 8vo. 14s.
Verney's Chess Eccentricities. Crown 8vo. 10s. 6d.
Wilcocks's Sea-Fisherman. Post 8vo. 6s.

ENCYCLOPÆDIAS, DICTIONARIES, AND BOOKS OF REFERENCE.

Acton's Modern Cookery for Private Families. Fcp. 8vo. 4s. 6d.
Ayre's Treasury of Bible Knowledge. Fcp. 8vo. 6s.
Brande's Dictionary of Science, Literature, and Art. 3 vols. medium 8vo. 63s.
Cabinet Lawyer (The), a Popular Digest of the Laws of England. Fcp. 8vo. 9s.
Cates's Dictionary of General Biography. Medium 8vo. 28s.
Doyle's The Official Baronage of England. Vols. L.-III. 3 vols. 4to. £5. 5s.
Gwilt's Encyclopædia of Architecture. 8vo. 52s. 6d.
Keith Johnston's Dictionary of Geography, or General Gazetteer. 8vo. 42s.
M'Culloch's Dictionary of Commerce and Commercial Navigation. 8vo. 63s.
Maunder's Biographical Treasury. Fcp. 8vo. 6s.
 — Historical Treasury. Fcp. 8vo. 6s.
 — Scientific and Literary Treasury. Fcp. 8vo. 6s.
 — Treasury of Bible Knowledge, edited by Ayre. Fcp. 8vo. 6s.
 — Treasury of Botany, edited by Lindley & Moore. Two Parts, 12s.
 — Treasury of Geography. Fcp. 8vo. 6s.
 — Treasury of Knowledge and Library of Reference. Fcp. 8vo. 6s.
 — Treasury of Natural History. Fcp. 8vo. 6s.
Quain's Dictionary of Medicine. Medium 8vo. 31s. 6d., or in 2 vols. 34s.
Reeve's Cookery and Housekeeping. Crown 8vo. 7s. 6d.
Rich's Dictionary of Roman and Greek Antiquities. Crown 8vo. 7s. 6d.
Roget's Thesaurus of English Words and Phrases. Crown 8vo. 10s. 6d.
Ure's Dictionary of Arts, Manufactures, and Mines. 4 vols. medium 8vo. £7. 7s.
Willich's Popular Tables, by Marriott. Crown 8vo. 10s. 6d.

London: LONGMANS, GREEN, & CO.

A SELECTION
OF
EDUCATIONAL WORKS.

TEXT-BOOKS OF SCIENCE
FULLY ILLUSTRATED.

Abney's Treatise on Photography. Fcp. 8vo. 3s. 6d.
Anderson's Strength of Materials. 3s. 6d.
Armstrong's Organic Chemistry. 3s. 6d.
Ball's Elements of Astronomy. 6s.
Barry's Railway Appliances. 3s. 6d.
Bauerman's Systematic Mineralogy. 6s.
— Descriptive Mineralogy. 6s.
Bloxam and Huntington's Metals. 5s.
Glazebrook's Physical Optics. 6s.
Glazebrook and Shaw's Practical Physics. 6s.
Gore's Art of Electro-Metallurgy. 6s.
Griffin's Algebra and Trigonometry. 3s. 6d. Notes and Solutions, 3s. 6d.
Holmes's The Steam Engine. 6s.
Jenkin's Electricity and Magnetism. 3s. 6d.
Maxwell's Theory of Heat. 3s. 6d.
Merrifield's Technical Arithmetic and Mensuration. 3s. 6d. Key, 3s. 6d.
Miller's Inorganic Chemistry. 3s. 6d.
Preece and Sivewright's Telegraphy. 5s.
Rutley's Study of Rocks, a Text-Book of Petrology. 4s. 6d.
Shelley's Workshop Appliances. 4s. 6d.
Thomé's Structural and Physiological Botany. 6s.
Thorpe's Quantitative Chemical Analysis. 4s. 6d.
Thorpe and Muir's Qualitative Analysis. 3s. 6d.
Tilden's Chemical Philosophy. 3s. 6d. With Answers to Problems. 4s. 6d.
Unwin's Elements of Machine Design. 6s.
Watson's Plane and Solid Geometry. 3s. 6d.

THE GREEK LANGUAGE.

Bloomfield's College and School Greek Testament. Fcp. 8vo. 5s.
Bolland & Lang's Politics of Aristotle. Post 8vo. 7s. 6d.
Collis's Chief Tenses of the Greek Irregular Verbs. 8vo. 1s.
— Pontes Graeci, Stepping-Stone to Greek Grammar. 12mo. 3s. 6d.
— Praxis Graeca, Etymology. 12mo. 2s. 6d.
— Greek Verse-Book, Praxis Iambica. 12mo. 4s. 6d.
Farrar's Brief Greek Syntax and Accidence. 12mo. 4s. 6d.
— Greek Grammar Rules for Harrow School. 12mo. 1s. 6d.
Geare's Notes on Thucydides. Book I. Fcp. 8vo. 2s. 6d.
Hewitt's Greek Examination-Papers. 12mo. 1s. 6d.
Isbister's Xenophon's Anabasis, Books I. to III. with Notes. 12mo. 3s. 6d.
Jerram's Graece Reddenda. Crown 8vo. 1s. 6d.

London: LONGMANS, GREEN, & CO.

14   A Selection of Educational Works.

Kennedy's Greek Grammar. 12mo. 4s. 6d.
Liddell & Scott's English-Greek Lexicon. 4to. 36s.; Square 12mo. 7s. 6d.
Mahaffy's Classical Greek Literature. Crown 8vo. Poets, 7s. 6d. Prose Writers, 7s. 6d.
Morris's Greek Lessons. Square 18mo. Part I. 2s. 6d.; Part II. 1s.
Parry's Elementary Greek Grammar. 12mo. 3s. 6d.
Plato's Republic, Book I. Greek Text, English Notes by Hardy. Crown 8vo. 3s.
Sheppard and Evans's Notes on Thucydides. Crown 8vo. 7s. 6d.
Thucydides, Book IV. with Notes by Barton and Chavasse. Crown 8vo. 5s.
Valpy's Greek Delectus, improved by White. 12mo. 2s. 6d. Key, 2s. 6d.
White's Xenophon's Expedition of Cyrus, with English Notes. 12mo. 7s. 6d.
Wilkins's Manual of Greek Prose Composition. Crown 8vo. 5s. Key, 5s.
— Exercises in Greek Prose Composition. Crown 8vo. 4s. 6d. Key, 2s. 6d.
— New Greek Delectus. Crown 8vo. 3s. 6d. Key, 2s. 6d.
— Progressive Greek Delectus. 12mo. 4s. Key, 2s. 6d.
— Progressive Greek Anthology. 12mo. 5s.
— Scriptores Attici, Excerpts with English Notes. Crown 8vo. 7s. 6d.
— Speeches from Thucydides translated. Post 8vo. 6s.
Yonge's English-Greek Lexicon. 4to. 21s.; Square 12mo. 8s. 6d.

THE LATIN LANGUAGE.

Bradley's Latin Prose Exercises. 12mo. 3s. 6d. Key, 5s.
— Continuous Lessons in Latin Prose. 12mo. 5s. Key, 5s. 6d.
— Cornelius Nepos, improved by White. 12mo. 3s. 6d.
— Eutropius, improved by White. 12mo. 2s. 6d.
— Ovid's Metamorphoses, improved by White. 12mo. 4s. 6d.
— Select Fables of Phædrus, improved by White. 12mo. 2s. 6d.
Collis's Chief Tenses of Latin Irregular Verbs. 8vo. 1s.
— Pontes Latini, Stepping-Stone to Latin Grammar. 12mo. 3s. 6d.
Hewitt's Latin Examination-Papers. 12mo. 1s. 6d.
Isbister's Cæsar, Books I.-VII. 12mo. 4s.; or with Reading Lessons, 4s. 6d.
— Cæsar's Commentaries, Books I.-V. 12mo. 3s. 6d.
— First Book of Cæsar's Gallic War. 12mo. 1s. 6d.
Jerram's Latiné Reddenda. Crown 8vo. 1s. 6d.
Kennedy's Child's Latin Primer, or First Latin Lessons. 12mo. 2s.
— Child's Latin Accidence. 12mo. 1s.
— Elementary Latin Grammar. 12mo. 3s. 6d.
— Elementary Latin Reading Book, or Tirocinium Latinum. 12mo. 2s.
— Latin Prose, Palæstra Stili Latini. 12mo. 6s.
— Latin Vocabulary. 12mo. 2s. 6d.
— Subsidia Primaria, Exercise Books to the Public School Latin Primer. I. Accidence and Simple Construction, 2s. 6d. II. Syntax, 3s. 6d.
— Key to the Exercises in Subsidia Primaria, Parts I. and II. price 5s.
— Subsidia Primaria, III. the Latin Compound Sentence. 12mo. 1s.
— Curriculum Stili Latini. 12mo. 4s. 6d. Key, 7s. 6d.
— Palæstra Latina, or Second Latin Reading Book. 12mo. 5s.

London: LONGMANS, GREEN, & CO.

A Selection of Educational Works. 15

Millington's Latin Prose Composition. Crown 8vo. 2s. 6d.
— Selections from Latin Prose. Crown 8vo. 2s. 6d.
Moody's Eton Latin Grammar. 12mo. 2s. 6d. The Accidence separately, 1s.
Morris's Elementa Latina. Fcp. 8vo. 1s. 6d. Key, 2s. 6d.
Parry's Origines Romanæ, from Livy, with English Notes. Crown 8vo. 4s.
The Public School Latin Primer. 12mo. 2s. 6d.
— — — — Grammar, by Rev. Dr. Kennedy. Post 8vo. 7s. 6d.
Prendergast's Mastery Series, Manual of Latin. 12mo. 2s. 6d.
Rapier's Introduction to Composition of Latin Verse. 12mo. 2s. 6d. Key, 2s. 6d.
Sheppard and Turner's Aids to Classical Study. 12mo. 5s. Key, 6s.
Valpy's Latin Delectus, improved by White. 12mo. 2s. 6d. Key, 2s. 6d.
Virgil's Æneid, translated into English Verse by Conington. Crown 8vo. 9s.
— Works, edited by Kennedy. Crown 8vo. 10s. 6d.
— — translated into English Prose by Conington. Crown 8vo. 9s.
Walford's Progressive Exercises in Latin Elegiac Verse. 12mo. 2s. 6d. Key, 5s.
White and Riddle's Large Latin-English Dictionary. 1 vol. 4to. 21s.
White's Concise Latin-Eng. Dictionary for University Students. Royal 8vo. 12s.
— Junior Students' Eng.-Lat. & Lat.-Eng. Dictionary. Square 12mo. 5s.
Separately { The Latin-English Dictionary, price 3s.
{ The English-Latin Dictionary, price 3s.
Yonge's Latin Gradus. Post 8vo. 9s.; or with Appendix, 12s.

WHITE'S GRAMMAR-SCHOOL GREEK TEXTS.

Æsop (Fables) & Palæphatus (Myths). 32mo. 1s.
Euripides, Hecuba. 2s.
Homer, Iliad, Book I. 1s.
— Odyssey, Book I. 1s.
Lucian, Select Dialogues. 1s.
Xenophon, Anabasis, Books I. III. IV. V. & VI. 1s. 6d. each; Book II. 1s.; Book VII. 2s.

Xenophon, Book I. without Vocabulary. 3d.
St. Matthew's and St. Luke's Gospels. 2s. 6d. each.
St. Mark's and St. John's Gospels. 1s. 6d. each.
The Acts of the Apostles. 2s. 6d.
St. Paul's Epistle to the Romans. 1s. 6d.

The Four Gospels in Greek, with Greek-English Lexicon. Edited by John T. White, D.D. Oxon. Square 32mo. price 5s.

WHITE'S GRAMMAR-SCHOOL LATIN TEXTS.

Cæsar, Gallic War, Books I. & II. V. & VI. 1s. each. Book I. without Vocabulary, 3d.
Cæsar, Gallic War, Books III. & IV. 9d. each.
Cæsar, Gallic War, Book VII. 1s. 6d.
Cicero, Cato Major (Old Age). 1s. 6d.
Cicero, Lælius (Friendship). 1s. 6d.
Eutropius, Roman History, Books I. & II. 1s. Books III. & IV. 1s.
Horace, Odes, Books I. II. & IV. 1s. each.
Horace, Odes, Book III. 1s. 6d.
Horace, Epodes and Carmen Seculare. 1s.

Nepos, Miltiades, Simon, Pausanias, Aristides. 9d.
Ovid, Selections from Epistles and Fasti. 1s.
Ovid, Select Myths from Metamorphoses. 9d.
Phædrus, Select Easy Fables.
Phædrus, Fables, Books I. & II. 1s.
Sallust, Bellum Catilinarium. 1s. 6d.
Virgil, Georgics, Book IV. 1s.
Virgil, Æneid. Books I. to VI. 1s. each. Book I. without Vocabulary, 3d.
Virgil, Æneid, Books VII. VIII. X. XI. XII. 1s. 6d. each.

London: LONGMANS, GREEN, & CO.

16   A Selection of Educational Works.

THE FRENCH LANGUAGE.

Albités's How to Speak French. Fcp. 8vo. 5s. 6d.
— Instantaneous French Exercises. Fcp. 2s. Key, 2s.
Cassal's French Genders. Crown 8vo. 3s. 6d.
Cassal & Karcher's Graduated French Translation Book.   Part I. 3s. 6d.
  Part II. 5s. Key to Part I. by Professor Cassal, price 5s.
Contanseau's Practical French and English Dictionary. Post 8vo. 3s. 6d.
— Pocket French and English Dictionary. Square 18mo. 1s. 6d.
— Premières Lectures. 12mo. 2s. 6d.
— First Step in French. 12mo. 2s. 6d. Key, 3s.
— French Accidence. 12mo. 2s. 6d.
— — Grammar. 12mo. 4s. Key, 3s.
Contanseau's Middle-Class French Course. Fcp. 8vo. :—

Accidence, 8d.   French Translation-Book, 8d.
Syntax, 8d.   Easy French Delectus, 8d.
French Conversation-Book, 8d.   First French Reader, 8d.
First French Exercise-Book, 8d.   Second French Reader, 8d.
Second French Exercise-Book, 8d.   French and English Dialogues, 8d.

Contanseau's Guide to French Translation. 12mo. 3s. 6d. Key 3s. 6d.
— Prosateurs et Poètes Français. 12mo. 5s.
— Précis de la Littérature Française. 12mo. 3s. 6d.
— Abrégé de l'Histoire de France. 12mo. 2s. 6d.
Féval's Chouans et Bleus, with Notes by C. Sankey, M.A. Fcp. 8vo. 2s. 6d.
Jerram's Sentences for Translation into French. Cr. 8vo. 1s. Key, 2s. 6d.
Prendergast's Mastery Series, French. 12mo. 2s. 6d.
Souvestre's Philosophe sous les Toits, by Stiévenard. Square 18mo. 1s. 6d.
Stepping-Stone to French Pronunciation. 18mo. 1s.
Stiévenard's Lectures Françaises from Modern Authors. 12mo. 4s. 6d.
— Rules and Exercises on the French Language. 12mo. 3s. 6d.
Tarver's Eton French Grammar. 12mo. 6s. 6d.

THE GERMAN LANGUAGE.

Blackley's Practical German and English Dictionary. Post 8vo. 3s. 6d.
Buchheim's German Poetry, for Repetition. 18mo. 1s. 6d.
Collis's Card of German Irregular Verbs. 8vo. 2s.
Fischer-Fischart's Elementary German Grammar. Fcp. 8vo. 2s. 6d.
Just's German Grammar. 12mo. 1s. 6d.
— German Reading Book. 12mo. 3s. 6d.
Longman's Pocket German and English Dictionary. Square 18mo. 2s. 6d.
Naftel's Elementary German Course for Public Schools. Fcp. 8vo.

German Accidence. 9d.   German Prose Composition Book. 9d.
German Syntax. 9d.   First German Reader. 9d.
First German Exercise-Book. 9d. Second German Reader. 9d.
Second German Exercise-Book. 9d.

Prendergast's Mastery Series, German. 12mo. 2s. 6d.
Quick's Essentials of German. Crown 8vo. 3s. 6d.
Selss's School Edition of Goethe's Faust. Crown 8vo. 5s.
— Outline of German Literature. Crown 8vo. 4s. 6d.
Wirth's German Chit-Chat. Crown 8vo. 2s. 6d.

London : LONGMANS, GREEN, & CO.

Spottiswoode & Co. Printers, New-street Square, London.

www.ingramcontent.com/pod-product-compliance
Lightning Source LLC
Chambersburg PA
CBHW020244240426
43672CD00006D/631